T0363785

SOLUTION MANUAL FOR

# CLASSICAL MECHANICS AND ELECTRODYNAMICS

**Second Edition**

SOLUTION MANUAL FOR

# CLASSICAL MECHANICS AND ELECTRODYNAMICS

## Second Edition

**Jon Magne Leinaas**
University of Oslo, Norway

NEW JERSEY • LONDON • SINGAPORE • BEIJING • SHANGHAI • HONG KONG • TAIPEI • CHENNAI

*Published by*

World Scientific Publishing Co. Pte. Ltd.
5 Toh Tuck Link, Singapore 596224
*USA office:* 27 Warren Street, Suite 401-402, Hackensack, NJ 07601
*UK office:* 57 Shelton Street, Covent Garden, London WC2H 9HE

Library of Congress Control Number: 2024950797

**British Library Cataloguing-in-Publication Data**
A catalogue record for this book is available from the British Library.

**SOLUTION MANUAL FOR CLASSICAL MECHANICS AND ELECTRODYNAMICS**
**Second Edition**



ISBN 978-981-98-0752-9 (hardcover)
ISBN 978-981-98-0665-2 (paperback)
ISBN 978-981-98-0666-9 (ebook for institutions)
ISBN 978-981-98-0667-6 (ebook for individuals)

For any available supplementary material, please visit
https://www.worldscientific.com/worldscibooks/10.1142/14144#t=suppl

Desk Editor: Muhammad Ihsan Putra

Typeset by Stallion Press
Email: enquiries@stallionpress.com

# Contents

# PART 1
# Analytical Mechanics

Chapter 1

# Generalized coordinates

## *Problem 1.1*

Four mechanical systems are studied. In all cases the number of degrees of freedom are specified, and an appropriate set of generalized coordinates is chosen.

a) The first system consists of a pendulum attached to a block which in turn is attached to a spring. We assume all motion takes place in a two-dimensional, vertical plane. The block is constrained to move in the horizontal direction, and the pendulum is constrained by the constant length of the rod. Starting from two degrees of freedom for each of the two objects, the two constraints reduce the number of degrees of freedom to two, one for each object. A natural choice of generalized coordinates is the horizontal displacement $x$ of the block and the angle $\theta$ of the rod relative to the vertical direction.

b) The second system consists of a pendulum attached to a vertical disk, which rotates with a fixed angular frequency. Also here we consider the motion restricted to a two-dimensional, vertical plane. There is no degree of freedom related to the rotating disk, since it has an externally determined angular frequency. The pendulum is again only constrained by the fixed length of the rod, and the number of degrees of freedom of the system is therefore one. A natural choice of generalized coordinate is the angle $\theta$ between the pendulum rod and the vertical direction.

c) In the third case a rigid rod can tilt without sliding on the top of the cylinder, while the cylinder can roll on a horizontal plane. Assuming again that the motion is restricted to a two-dimensional, vertical plane, the starting point is three degrees of freedom for each object. For the cylinder this corresponds to two coordinates for its center of mass and one for its angle of rotation. For the rod there are two coordinates needed to determine the position of its center of mass, and one coordinate to determine the angle of the rod relative to the horizontal (or vertical) direction.

The constraints are the following,
1) For the cylinder the vertical coordinate of the cylinder's center is fixed, and since the cylinder is rolling, rather than sliding, the rotation coordinate is linked to the horizontal coordinate of the cylinder. This gives two constraints for the cylinder.
2) The rod is constrained to lie on the top of the cylinder, and if we assume that it is not allowed to slide on the cylinder, the only degree of freedom for the rod is to tilt on the cylinder. This means that there are two constraints also for the rod.
The number of degrees of freedom is therefore: $3+3-2-2=2$. A possible choice of generalized coordinates is the horizontal coordinate of the cylinder and the tilting angle of the rod.

d) In the last case a rotating top moves on a horizontal floor. For a rigid body in three dimensions the number of degrees of freedom is six, three to determine the position of its center of mass and three to determine its orientation (corresponding to the three parameters, which are needed to specify a rotation in 3D). For the rotating top there is one constraint, with the vertical coordinate of its tip being fixed by the vertical coordinate of the floor. Choice of generalized coordinates: The $(x, y)$-coordinates of the tip of the rotating top, and the angles $(\theta, \phi, \chi)$ which determine its orientation (two to determine the direction of its symmetry axis, and one to determine its rotation angle around the axis).

### *Problem 1.2*

We examine an Atwood's machine, which consists of three weights, with masses $m_1 = 4m$, $m_2 = 2m$ and $m_3 = m$. The ropes, with fixed lengths $l_1$ and $l_2$, and the two pulleys, are treated as massless.

The number of degrees of freedom is two, since the heights of two of the weights, $m_1$ and $m_2$, will determine the third one, $m_3$. We choose the vertical coordinates $y_1$ and $y_2$, shown in Fig. 1.1, as generalized coordinates. Expressed in these variables the potential energy is

$$\begin{aligned}
V &= -m_1 y_1 + m_2(-y_2 - (l_1 - y_1)) + m_3((y_1 - l_1) + (y_2 - l_2)) \\
&= -m_1 y_1 + m_2(y_1 - y_2) + m_3(y_1 + y_2) + const. \\
&= -4m y_1 + 2m(y_1 - y_2) + m(y_1 + y_2) + const. \\
&= -m(y_1 + y_2) + const., \qquad (1.1)
\end{aligned}$$

where the constant can be absorbed in the definition of the zero point of the potential energy.

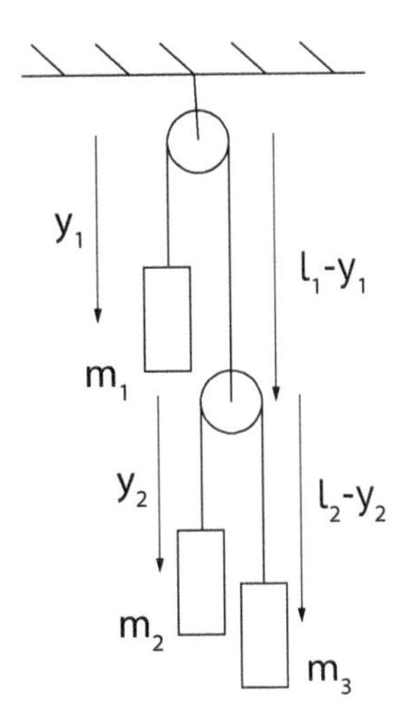


Fig. 1.1 Atwood's machine with two independent coordinates $y_1$ and $y_2$.

The corresponding expression for the kinetic energy is

$$
\begin{aligned}
T &= \frac{1}{2}m_1\dot{y}_1^2 + \frac{1}{2}m_2(\dot{y}_1 - \dot{y}_2)^2 + \frac{1}{2}m_3(\dot{y}_1 + \dot{y}_2)^2 \\
&= \frac{1}{2}m(4\dot{y}_1^2 + 2(\dot{y}_1 - \dot{y}_2)^2 + (\dot{y}_1 + \dot{y}_2)^2) \\
&= \frac{1}{2}m(7\dot{y}_1^2 + 3\dot{y}_2^2 - 2\dot{y}_1\dot{y}_2)\,. \qquad (1.2)
\end{aligned}
$$

### *Problem 1.3*

Three identical rods of mass $m$ and length $l$ are connected by frictionless joints, with the distance between the points of suspension being equal to the length of the rods. The rods move in the vertical plane. We will show that the system has only one degree of freedom, where the angle $\theta$ of one of the rods can be used as generalized coordinate. The Lagrangian will be found, expressed as a function of $\theta$ and $\dot{\theta}$.

Two coordinates are needed to determine the position of each of the joints in the vertical plane. There are three constraints, which give relations between these positions, corresponding to the fixed lengths of the three rods. The number of degrees of freedom of the system is therefore $2+2-3=1$, and the angle $\theta$ in the figure is an obvious choice for the generalized coordinate.

The four endpoints of the rods define a parallelogram, as follows from the fact that the lengths of all four sides are the same. As a consequence two of the rods will rotate about the end points, which are fixed to the roof, while the third rod will move without rotation, since it is always parallel to the roof.

The common angular velocity of the rotating rods is $\omega = \dot{\theta}$, and the velocity of the non-rotating rod is $v = l\dot{\theta}$, since this is equal to the velocity of the lower endpoints of the two rotating rods. The kinetic energy of the three rods is therefore

$$T = 2 \cdot \frac{1}{2} I\omega^2 + \frac{1}{2} mv^2 = \frac{1}{3} ml^2\dot{\theta}^2 + \frac{1}{2} ml^2\dot{\theta}^2 = \frac{5}{6} ml^2\dot{\theta}^2 \,. \tag{1.3}$$

With $y$ as the vertical distance of the horizontal rod, the potential energy of the system is

$$V = 2 \cdot mg\frac{y}{2} + mgy = 2mgy = -2mgl\cos\theta \,. \tag{1.4}$$

This gives for the Lagrangian

$$L(\theta, \dot{\theta}) = \frac{5}{6} ml^2\dot{\theta}^2 + 2mgl\cos\theta \,. \tag{1.5}$$

### *Problem 1.4*

A particle with mass $m$ moves in three-dimensional space under the influence of a constraint. The constraint is expressed by the following relation between the Cartesian coordinates of the particle,

$$e^{-(x^2+y^2)} + z = 0 \,. \tag{1.6}$$

a) The constraint relation can be used to express $z$ as a function of $x$ and $y$. Thus, there are two independent variables, which means that the system has two degrees of freedom. With $x$ and $y$ chosen as the generalized coordinates, the position vector of the particle is

$$\mathbf{r} = x\mathbf{i} + y\mathbf{j} - e^{-(x^2+y^2)}\mathbf{k} \,. \tag{1.7}$$

b) A virtual displacement, with $x \to x + \delta x$ and $y \to y + \delta y$, gives the following variation in $z$

$$\delta z = 2e^{-(x^2+y^2)}(x\delta x + y\delta y) \,. \tag{1.8}$$

The expression for the variation of the position vector is therefore

$$\delta\mathbf{r} = (\mathbf{i} + 2e^{-(x^2+y^2)}x\,\mathbf{k})\delta x + (\mathbf{j} + 2e^{-(x^2+y^2)}y\,\mathbf{k})\delta y \,. \tag{1.9}$$

c) The constraint force $\mathbf{f}$ satisfies the condition $\mathbf{f} \cdot \delta\mathbf{r} = 0$ for arbitrary variations of the form (1.9). This implies that the two coefficients of the scalar product, proportional to $\delta x$ and $\delta y$ respectively, have to vanish separately,

$$f_x + 2e^{-(x^2+y^2)}xf_z = 0\,, \quad f_y + 2e^{-(x^2+y^2)}yf_z = 0. \tag{1.10}$$

In vector form this is

$$\mathbf{f} = f_z(-2e^{-(x^2+y^2)}x\mathbf{i} - 2e^{-(x^2+y^2)}y\mathbf{j} + \mathbf{k})\,. \tag{1.11}$$

Only the direction of $\mathbf{f}$ is determined by this expression since $f_z$ is an undetermined function of $x$ and $y$. To fully determine the constraint force which acts on the particle one needs to know the applied forces and the velocity of the particle.

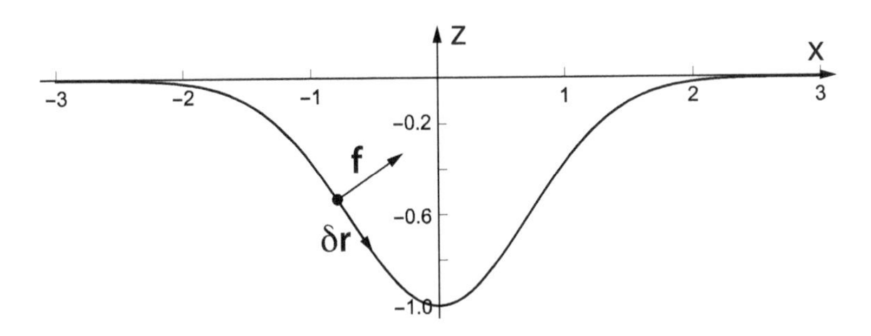


Fig. 1.2 A section of the constraint surface with directions of a virtual displacement $\delta\mathbf{r}$ of the small body and the constraint force $\mathbf{f}$ acting on the body.

d) In the $y = 0$ plane the constraint is described by the curve $z = -e^{-x^2}$. A virtual displacement $\delta\mathbf{r}$ is directed tangentially to the curve, while the constraint force $\mathbf{f}$ is perpendicular to the curve, as illustrated in Fig 1.2.

### *Problem 1.5*

A flexible chain can move without friction on a smooth surface with the vertical heights of the endpoints denoted $z_A$ and $z_B$. The chain has constant (linear) mass density $\mu$. We shall use the principle of virtual work to find how $z_A$ and $z_B$ are related when the chain is at static equilibrium.

A (dimensionless) parameter $s$, which measures length along the chain (relative to the total length), will be used as the variable. We set $s = 0$ at the right end of the chain and $s = 1$ at the left end. For a virtual translation

$\delta s$ of the chain along the surface, there is a vertical displacement $\delta z$, which varies with the position $s$ along the chain,

$$\delta z = \frac{dz}{ds}\delta s\,. \tag{1.12}$$

The corresponding virtual work, integrated along the chain is

$$\delta W = g\int \delta z dm = g\mu\delta s\int_0^1 \frac{dz}{ds}ds = g\mu\delta s(z_B - z_A)\,. \tag{1.13}$$

At equilibrium the principle of virtual work gives $\delta W = 0$, which implies $z_A = z_B$.

Note that, as a simpler argument, a small displacement of the chain along the surface is equivalent to taking a small part of the chain at one end and move it to the other end. At equilibrium the change in potential energy should vanish, which would again mean that $z_A = z_B$.

Chapter 2

# Lagrange's equations

*Problem 2.1*

A particle with mass $m$ moves freely in a horizontal plane. The problem to be solved is to give a Lagrangian description of the motion in a rotating coordinate system, and to compare the corresponding equations of motion with the standard Newtonian description, where the rotation introduces centrifugal and Coriolis forces.

a) The coordinate transformation from the rotating Cartesian reference frame, with coordinates $(\xi, \eta)$, to the fixed Cartesian frame, with coordinates $(x, y)$, is

$$\begin{aligned} x &= \xi \cos \omega t - \eta \sin \omega t\,, \\ y &= \xi \sin \omega t + \eta \cos \omega t\,, \end{aligned} \tag{2.1}$$

with the corresponding transformation of velocities,

$$\begin{aligned} \dot{x} &= (\dot{\xi} - \omega \eta) \cos \omega t - (\omega \xi + \dot{\eta}) \sin \omega t\,, \\ \dot{y} &= (\dot{\xi} - \omega \eta) \sin \omega t + (\omega \xi + \dot{\eta}) \cos \omega t\,. \end{aligned} \tag{2.2}$$

The Lagrangian is identical to the kinetic energy,

$$\begin{aligned} L &= \frac{1}{2} m (\dot{x}^2 + \dot{y}^2) \\ &= \frac{1}{2} m \left[ (\dot{\xi} - \omega \eta)^2 + (\omega \xi + \dot{\eta})^2 \right] \\ &= \frac{1}{2} m \left[ (\dot{\xi}^2 + \dot{\eta}^2) + 2\omega (\xi \dot{\eta} - \dot{\xi} \eta) + \omega^2 (\xi^2 + \eta^2) \right]\,. \end{aligned} \tag{2.3}$$

b) The partial derivatives of L with respect to the coordinates and their time derivative are

$$\begin{aligned} \frac{\partial L}{\partial \xi} &= m(\omega \dot{\eta} + \omega^2 \xi)\,, \quad & \frac{\partial L}{\partial \dot{\xi}} &= m(\dot{\xi} - \omega \eta)\,, \\ \frac{\partial L}{\partial \eta} &= m(-\omega \dot{\xi} + \omega^2 \eta)\,, \quad & \frac{\partial L}{\partial \dot{\eta}} &= m(\dot{\eta} + \omega \xi)\,. \end{aligned} \tag{2.4}$$

This gives as Lagrange's equations,

$$\begin{aligned}
\frac{d}{dt}\frac{\partial L}{\partial \dot{\xi}} - \frac{\partial L}{\partial \xi} = 0 \quad &\Rightarrow \quad \ddot{\xi} - 2\omega\dot{\eta} - \omega^2\xi = 0\,, \\
\frac{d}{dt}\frac{\partial L}{\partial \dot{\eta}} - \frac{\partial L}{\partial \eta} = 0 \quad &\Rightarrow \quad \ddot{\eta} + 2\omega\dot{\xi} - \omega^2\eta = 0\,.
\end{aligned} \tag{2.5}$$

Without any external force acting on the particle, Newton's second law takes in the rotating frame the following form

$$m\ddot{\mathbf{r}} = -m\boldsymbol{\omega} \times (\boldsymbol{\omega} \times \mathbf{r}) - 2m\boldsymbol{\omega} \times \dot{\mathbf{r}}\,, \tag{2.6}$$

where the angular velocity vector $\boldsymbol{\omega}$ is orthogonal to the plane of rotation. The term which is linear in $\omega$ is the Coriolis force and the term which is quadratic in $\omega$ is the centrifugal force. To compare the equations in (2.5) and (2.6) we express the position vector $\mathbf{r}$ in (2.6) in terms of the coordinates $\xi$ and $\eta$ as

$$\mathbf{r} = \xi\mathbf{i}' + \eta\mathbf{j}'\,, \tag{2.7}$$

where $\mathbf{i}'$ and $\mathbf{j}'$ are rotating unit vectors. In the rotating frame these are treated as fixed and the velocity and acceleration vectors are therefore given as

$$\dot{\mathbf{r}} = \dot{\xi}\mathbf{i}' + \dot{\eta}\mathbf{j}'\,, \quad \ddot{\mathbf{r}} = \ddot{\xi}\mathbf{i}' + \ddot{\eta}\mathbf{j}'\,. \tag{2.8}$$

We insert these expressions for the Coriolis and centrifugal forces

$$\begin{aligned}
\boldsymbol{\omega} \times (\boldsymbol{\omega} \times \mathbf{r}) = \boldsymbol{\omega}(\boldsymbol{\omega} \cdot \mathbf{r}) - \omega^2\mathbf{r} &= -\omega^2(\xi\mathbf{i}' + \eta\mathbf{j}')\,, \\
\boldsymbol{\omega} \times \dot{\mathbf{r}} = \omega\mathbf{k} \times (\dot{\xi}\mathbf{i}' + \dot{\eta}\mathbf{j}') &= \omega(\dot{\xi}\mathbf{j}' - \dot{\eta}\mathbf{i}')\,.
\end{aligned} \tag{2.9}$$

Inserting these in the vector equation (2.6) and extracting the components proportional to $\mathbf{i}'$ and $\mathbf{j}'$, we find

$$\begin{aligned}
\ddot{\xi} &= 2\omega\dot{\eta} + \omega^2\xi\,, \\
\ddot{\eta} &= -2\omega\dot{\xi} + \omega^2\eta\,,
\end{aligned} \tag{2.10}$$

which are the same two equations as in (2.5). This shows the consistency between Lagrange's equations expressed in the (rotating) coordinates $\xi$ and $\eta$, and the standard vector equation used for Newton's second law in a rotating reference frame.

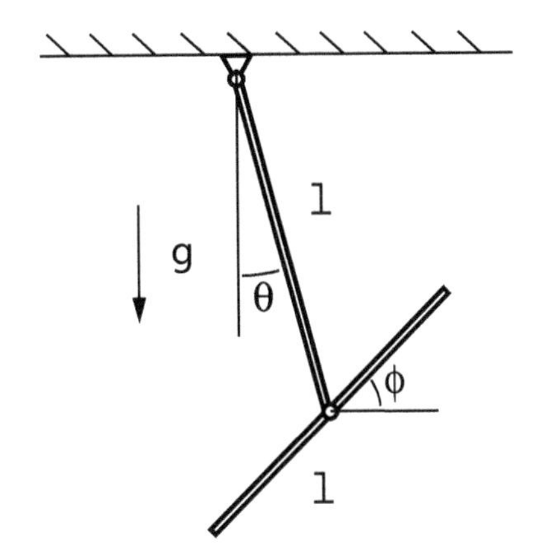


Fig. 2.1 The two-rod problem, with generalized coordinates $\theta$ and $\phi$.

***Problem 2.2***

The system consists here of two identical rods of mass $m$ and length $l$, which are connected to each other by a frictionless joint, as shown in Fig. 2.1. The problem is to give a Lagrangian description of the system and to find the angular frequency for small oscillations about its equilibrium position.

a) The system has two degrees of freedom, and as generalized coordinates we may choose one angle for each of the rods (see Fig. 2.1). With $I_1$ as the moment of inertia of the upper rod about its fixed endpoint, and $I_2$ as the moment of inertia of the lower rod about its middle point, the kinetic energy of the system is

$$\begin{aligned} T &= \frac{1}{2}I_1\dot{\theta}^2 + \frac{1}{2}ml^2\dot{\theta}^2 + \frac{1}{2}I_2\dot{\phi}^2 \\ &= \frac{1}{6}ml^2\dot{\theta}^2 + \frac{1}{2}ml^2\dot{\theta}^2 + \frac{1}{24}ml^2\dot{\phi}^2 \\ &= \frac{2}{3}ml^2\dot{\theta}^2 + \frac{1}{24}ml^2\dot{\phi}^2\,, \end{aligned} \tag{2.11}$$

and the potential energy is

$$\begin{aligned} V &= -mg\frac{l}{2}\cos\theta - mgl\cos\theta \\ &= -\frac{3}{2}mgl\cos\theta\,. \end{aligned} \tag{2.12}$$

This gives the Lagrangian $L = T - V$,

$$L = \frac{2}{3}ml^2\dot{\theta}^2 + \frac{1}{24}ml^2\dot{\phi}^2 + \frac{3}{2}mgl\cos\theta\,. \tag{2.13}$$

b) The derivatives with respect to $\theta$ and $\dot{\theta}$ are

$$\frac{\partial L}{\partial \dot{\theta}} = \frac{4}{3} m l^2 \dot{\theta}\,, \quad \frac{\partial L}{\partial \theta} = -\frac{3}{2} m g l \sin\theta\,. \tag{2.14}$$

This gives Lagrange's equation for the variable $\theta$,

$$\frac{d}{dt}\frac{\partial L}{\partial \dot{\theta}} - \frac{\partial L}{\partial \dot{\theta}} = 0 \quad \Rightarrow \quad \ddot{\theta} + \frac{9g}{8l}\sin\theta = 0\,. \tag{2.15}$$

The derivatives with respect to $\phi$ and $\dot{\phi}$ are

$$\frac{dL}{d\dot{\phi}} = \frac{1}{12} m l^2 \dot{\phi}\,, \quad \frac{\partial L}{\partial \phi} = 0\,. \tag{2.16}$$

Since $\phi$ is cyclic, the derivative with respect to $\dot{\phi}$ is a constant, and therefore the angular velocity $\dot{\phi}$ is constant. Thus the motion of the two rods are independent, with the upper rods making oscillations about $\theta = 0$, while the lower rod is rotating with constant angular velocity. For small oscillations, with $\sin\theta \approx \theta$, the angular frequency of the upper rod is $\omega = \sqrt{\frac{9g}{8l}}$.

***Problem 2.3***

We consider a small body with mass $m$, which moves without friction along a rotating rod in the horizontal plane. The angular velocity $\omega$ of the rod is constant, and the center of rotation is assigned the radial coordinate $r = 0$. The problem is to use Lagrange's method to determine the time dependent coordinate $r(t)$ of the moving body, and to plot the orbit of the body in the plane.

a) The position vector of the body is

$$\mathbf{r} = r(\cos\omega t\mathbf{i} + \sin\omega t\mathbf{j})\,, \tag{2.17}$$

which gives the velocity vector

$$\dot{\mathbf{r}} = -r\omega(\sin\omega t\mathbf{i} - \cos\omega t\mathbf{j}) + \dot{r}(\cos\omega t\mathbf{i} + \sin\omega t\mathbf{j})\,. \tag{2.18}$$

Since the body moves in the horizontal plane there is no potential energy, and the Lagrangian is identical to the kinetic energy

$$L = \frac{1}{2} m \dot{\mathbf{r}}^2 = \frac{1}{2} m(\dot{r}^2 + r^2\omega^2)\,. \tag{2.19}$$

The partial derivatives with respect to $r$ and $\dot{r}$ are

$$\frac{\partial L}{\partial \dot{r}} = m\dot{r}\,, \quad \frac{\partial L}{\partial r} = m r \omega^2\,, \tag{2.20}$$

and Lagrange's equation gives

$$\frac{d}{dt}\frac{\partial L}{\partial \dot{r}} - \frac{\partial L}{\partial r} = 0 \quad \Rightarrow \quad \ddot{r} - \omega^2 r = 0\,. \tag{2.21}$$

Solutions of the equations are

$$r(t) = Ae^{\omega t} + Be^{-\omega t}\,. \tag{2.22}$$

The initial conditions are $r(0) = r_0$ and $\dot{r}(0) = 0$, which give $A = B = \frac{1}{2}r_0$. This gives the solution

$$r(t) = \frac{1}{2}r_0(e^{\omega t} + e^{-\omega t}) = r_0 \cosh \omega t\,. \tag{2.23}$$

b) The figure shows a plot of the orbit in the $x, y$-plane, with $t$ restricted to the interval $0 < t \lesssim \pi/\omega$.

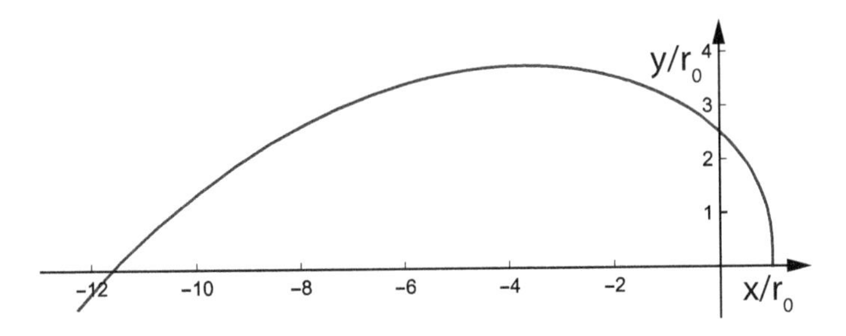


Fig. 2.2 The orbit of the body sliding on a rotating rod.

## *Problem 2.4*

A pendulum consists of a rigid rod of length $l$, which we consider as massless, and a pendulum bob of mass $m$. The point of suspension of the pendulum has horizontal coordinate $x = s$ and vertical coordinate $y = 0$. The angle $\theta$ of the pendulum rod, relative to the vertical direction, is used as generalized coordinate.

a) We assume first that $s = 0$. The coordinates and velocities of the pendulum bob are

$$\begin{aligned} x &= l\sin\theta\,, \quad y = -l\cos\theta\,, \\ \dot{x} &= l\cos\theta\,\dot{\theta}\,, \quad \dot{y} = l\sin\theta\,\dot{\theta}\,. \end{aligned} \tag{2.24}$$

The kinetic energy is

$$T = \frac{1}{2}m(\dot{x}^2 + \dot{y}^2) = \frac{1}{2}ml^2\dot{\theta}^2\,, \tag{2.25}$$

and the potential energy

$$V = mgy = -mgl\cos\theta\,. \tag{2.26}$$

This gives as Lagrangian $L = T - V$,

$$L = \frac{1}{2}ml^2\dot{\theta}^2 + mgl\cos\theta\,, \tag{2.27}$$

with partial derivatives

$$\frac{\partial L}{\partial\dot{\theta}} = ml^2\dot{\theta}\,, \quad \frac{\partial L}{\partial\theta} = -mgl\sin\theta\,, \tag{2.28}$$

and Lagrange's equation

$$\frac{d}{dt}\frac{\partial L}{\partial\dot{\theta}} - \frac{\partial L}{\partial\theta} = 0 \quad \Rightarrow \quad \ddot{\theta} + \frac{g}{l}\sin\theta = 0\,, \tag{2.29}$$

which is the standard pendulum equation.

b) With the point of suspension freely moving, and $s$ and $\theta$ as generalized coordinates, the expressions for $x$, $y$, and their time derivatives are

$$\begin{aligned} x &= s + l\sin\theta\,, \qquad y = -l\cos\theta, \\ \dot{x} &= \dot{s} + l\cos\theta\,\dot{\theta}\,, \quad \dot{y} = l\sin\theta\,\dot{\theta}\,. \end{aligned} \tag{2.30}$$

This gives for the kinetic energy

$$\begin{aligned} T &= \frac{1}{2}m(\dot{x}^2 + \dot{y}^2) \\ &= \frac{1}{2}ml^2\dot{\theta}^2 + ml\cos\theta\dot{s}\dot{\theta} + \frac{1}{2}m\dot{s}^2\,, \end{aligned} \tag{2.31}$$

while the potential energy is unchanged from (2.26). The Lagrangian is then

$$L = \frac{1}{2}ml^2\dot{\theta}^2 + ml\cos\theta\dot{s}\dot{\theta} + \frac{1}{2}m\dot{s}^2 + mgl\cos\theta\,, \tag{2.32}$$

with partial derivatives

$$\begin{aligned} \frac{\partial L}{\partial\dot{\theta}} &= ml^2\dot{\theta} + ml\cos\theta\dot{s}\,, \\ \frac{\partial L}{\partial\theta} &= -mgl\sin\theta - ml\sin\theta\dot{s}\dot{\theta}\,, \\ \frac{\partial L}{\partial\dot{s}} &= m\dot{s} + ml\cos\theta\dot{\theta}\,, \\ \frac{\partial L}{\partial s} &= 0\,. \end{aligned} \tag{2.33}$$

When introduced in Lagrange's equation, these expressions give rise to the following two equations of motion

$$\begin{aligned} l\ddot{\theta} + g\sin\theta + \cos\theta\ddot{s} &= 0, \\ \ddot{s} - l\sin\theta\dot{\theta}^2 + l\cos\theta\ddot{\theta} &= 0\,. \end{aligned} \tag{2.34}$$

c) The variable $s$ can be eliminated by a combination of the two equations, and this gives the following equation for the $\theta$ variable

$$l\sin^2\theta\ddot{\theta} + l\sin\theta\cos\theta\dot{\theta}^2 + g\sin\theta = 0\,. \tag{2.35}$$

A particular solution of this is $\theta = 0$, which when applied to (2.34) gives $\ddot{s} = 0$. This solution means that the pendulum rod is vertically oriented and moves with constant speed in the horizontal direction.

Under the assumption $\theta \neq 0$, Eq. (2.35) can be written as

$$\frac{d^2}{dt^2}(-l\cos\theta) = l(\sin\theta\ddot{\theta} + \cos\theta\dot{\theta}^2) = -g\,, \tag{2.36}$$

and since the vertical coordinate of the pendulum bob is $y = -l\cos\theta$, this means that the vertical acceleration of the bob is identical to the acceleration of gravity, $\ddot{y} = -g$.

We further note that the second equation in (2.34) can be written as

$$\ddot{s} = -\frac{d^2}{dt^2}(l\sin\theta). \tag{2.37}$$

Since the horizontal coordinate of the bob is $x = s + l\sin\theta$, this means that the acceleration in the horizontal direction vanishes, $\ddot{x} = 0$. Combining the results for the $x$ and $y$ variables, we conclude that the motion of the pendulum bob is like free fall in the gravitational field, in spite of the fact that there is a constraint on the motion of the upper end of the pendulum rod. However, there is a limit to this motion for $y = -l$ or $\theta = 0$, where the constraint will stop the downward motion.

### *Problem 2.5*

A rigid circular metal hoop rotates with constant angular velocity $\omega$ around an axis through the center. A bead with mass $m$ slides without friction along the circle and there is no gravity. We use the angular variable $\theta$ of the bead, measured around the hoop, as generalized coordinate. Lagrange's equation for the moving bead is established, and the angular frequency of small oscillations around stable equilibria is found.

a) The Cartesian coordinates of the particle are

$$x = R\sin\theta\cos\omega t\,, \quad y = R\sin\theta\sin\omega t\,, \quad z = -R\cos\theta. \tag{2.38}$$

Since there is no potential energy, the Lagrangian is given by the kinetic energy, $L = T$,

$$\begin{aligned} L &= \frac{1}{2}m(\dot{x}^2 + \dot{y}^2 + \dot{z}^2) \\ &= \frac{1}{2}mR^2[(\dot{\theta}\cos\theta\cos\omega t - \omega\sin\theta\sin\omega t)^2 \\ &\quad +(\dot{\theta}\cos\theta\sin\omega t + \omega\sin\theta\cos\omega t)^2 + \dot{\theta}^2\sin^2\theta \\ &= \frac{1}{2}mR^2(\dot{\theta}^2 + \omega^2\sin^2\theta)]\,. \end{aligned} \tag{2.39}$$

This gives Lagrange's equation,

$$mR^2(\ddot{\theta} - \omega^2\sin\theta\cos\theta) = 0 \quad \Rightarrow \quad \ddot{\theta} - \omega^2\sin\theta\cos\theta = 0\,. \tag{2.40}$$

b) Both terms in the Lagrangian comes from the kinetic energy of the particle, with no contribution from a potential energy. However, we see that the Lagrangian is identical to the Lagrangian for a particle with mass $m$ moving on a circle with radius $R$, where $T = \frac{1}{2}mR^2\dot{\theta}^2$ is the kinetic energy and $V = -\frac{1}{2}mR^2\omega^2\sin^2\theta$ is a periodic potential. The potential has two minima on the circle, with $\sin^2\theta = 1$ for $\theta = \pi/2$ and $\theta = 3\pi/2$. These correspond to stable equilibria. There are also two maxima, with $\sin^2\theta = 0$ for $\theta = 0$ and $\theta = \pi$, and these correspond to unstable equilibria. The potential $V$ can be viewed as a centrifugal potential, as described in a rotating reference frame which is co-rotating with the hoop.

c) With $\theta_0 = \pi/2$ we have for small deviations $\phi$ from this value,

$$\begin{aligned} \sin\theta &= \sin(\phi + \frac{\pi}{2}) = \cos\phi \approx 1, \\ \cos\theta &= \cos(\phi + \frac{\pi}{2}) = -\sin\phi \approx \phi. \end{aligned} \tag{2.41}$$

The equation of motion (2.40), to first order in the variable $\phi$, is then

$$\ddot{\phi} + \omega^2\phi = 0 \tag{2.42}$$

which is a harmonic oscillator equation with angular frequency $\omega$.

In the case of the second minimum, $\theta_0 = 3\pi/2$, we note that the shift $\theta \to \theta + \pi$ will only introduce a sign change for $\cos\theta$ and $\sin\theta$. The equation of motion (2.40) is therefore unchanged under this shift, and the small oscillation form of the equation therefore is the same for $\theta_0 = 3\pi/2$ as for $\theta_0 = \pi/2$.

*Problem 2.6*

We study the motion of an object with mass $m$, which slides without friction on an inclined plane. The plane, which is tilted with an angle of 30° relative to the horizontal plane, is forced to move horizontally with a constant acceleration $a$. The displacement $s$ of the object along the tilted surface is used as generalized coordinate.

a) We assume first that the inclined plane is at rest, $a = 0$. The Cartesian coordinates of the body are, expressed in terms of the parameter $s$,

$$x = s\cos 30^\circ = \frac{s}{2}\sqrt{3}\,, \quad y = h - s\sin 30^\circ = h - \frac{s}{2}. \tag{2.43}$$

The kinetic energy is then

$$T = \frac{1}{2}m(\dot{x}^2 + \dot{y}^2) = \frac{1}{2}m\dot{s}^2\,, \tag{2.44}$$

and the potential energy is

$$V = mgy = mg(h - \frac{s}{2})\,. \tag{2.45}$$

This gives the Lagrangian

$$L = \frac{1}{2}m\dot{s}^2 - mg(h - \frac{s}{2})\,. \tag{2.46}$$

b) We assume next that the acceleration $a$ of the inclined plane is constant and non-vanishing. The Cartesian coordinates, and their time derivatives are now

$$\begin{aligned} x &= \frac{1}{2}at^2 + \frac{s}{2}\sqrt{3}\,, \quad y = h - \frac{s}{2}\,, \\ \dot{x} &= at + \frac{\dot{s}}{2}\sqrt{3}\,, \quad \dot{y} = -\frac{\dot{s}}{2}\,. \end{aligned} \tag{2.47}$$

This gives the Lagrangian

$$L = \frac{1}{2}m(\dot{s}^2 + at\dot{s}\sqrt{3} + a^2t^2) - mg(h - \frac{s}{2})\,. \tag{2.48}$$

c) The partial derivatives with respect to $\dot{s}$ and $s$ are

$$\frac{\partial L}{\partial \dot{s}} = m(\dot{s} + \frac{1}{2}\sqrt{3}at)\,, \quad \frac{\partial L}{\partial s} = \frac{1}{2}mg\,, \tag{2.49}$$

which give Lagrange's equation

$$\frac{d}{dt}\frac{\partial L}{\partial \dot{s}} - \frac{\partial L}{\partial s} = 0 \quad \Rightarrow \quad \ddot{s} = \frac{1}{2}g - \frac{1}{2}\sqrt{3}\,a\,. \tag{2.50}$$

Since $\ddot{s}$ is constant, it is straight forward to integrate it to give $s$ as a function of $t$. With the initial conditions for $t = 0$, that $s = 0$ and $\dot{s} = 0$, the solution is

$$s(t) = \frac{1}{4}(g - a\sqrt{3})t^2\,. \tag{2.51}$$

### *Problem 2.7*

Two bodies with the same mass, $m$, are connected with a massless rope through a small hole in a smooth horizontal plane. One body is moving on the plane, the other one is hanging at the end of the rope and can move vertically. The polar coordinates $(r, \phi)$ of the body moving on the plane are used as generalized coordinates.

a) The kinetic energy of the system is

$$T = \frac{1}{2}m(\dot{r}^2 + r^2\dot{\phi}^2) + \frac{1}{2}m\dot{r}^2 = m(\dot{r}^2 + \frac{1}{2}r^2\dot{\phi}^2)\,, \tag{2.52}$$

and the potential energy is

$$V = -mg(l - r)\,, \tag{2.53}$$

with $l$ as the length of the rope. This gives as the Lagrangian $L = T - V$,

$$L = m(\dot{r}^2 + \frac{1}{2}r^2\dot{\phi}^2) - mgr + constant\,. \tag{2.54}$$

Lagrange's equation for the $r$ variable then is

$$\frac{d}{dt}\frac{\partial L}{\partial \dot{r}} - \frac{\partial L}{\partial r} = 0 \quad \Rightarrow \quad \ddot{r} - r\dot{\phi}^2 + g = 0\,. \tag{2.55}$$

b) The variable $\phi$ is cyclic, $\frac{\partial L}{\partial \phi} = 0$, which reduces Lagrange's equation for $\phi$ to

$$\frac{\partial L}{\partial \dot{\phi}} = mr^2\dot{\phi} = md\,, \tag{2.56}$$

with $d$ as a constant. The expression it gives for $\dot{\phi}$ can be used to eliminate $\dot{\phi}$ in the radial equation, which then takes the form

$$\ddot{r} - \frac{d^2}{r^3} + g = 0\,. \tag{2.57}$$

The equation shows that there is a special solution with constant $r$, with the value $r_0 = (d^2/(gl))^{1/3}$. For this value of $r$ the body on the horizontal plane will move with constant angular velocity along the circle with radius $r_0$. The equation of motion also shows that for $r > r_0$ $\ddot{r}$ will be negative, and for $r < r_0$ it will be positive. This means that the body will more generally oscillate about the circle $r = r_0$ under the angular motion.

***Problem 2.8***

A pendulum is connected to a block, which can slide without friction in a horizontal direction. The block and the pendulum bob have equal masses $m$, while the pendulum rod, with length $d$, is considered to be massless. As generalized coordinates in this problem, we use $s$ as the $x$-coordinate of the center of mass of the box, and $\theta$ as the angle of the pendulum rod relative to the vertical direction. At time $t = 0$ both the block and the pendulum have zero velocity, with the pendulum angle being $\theta_0$.

a) With $(x, y)$ as coordinates of the pendulum bob we have

$$x = s + d\sin\theta\,, \quad y = -d\cos\theta. \tag{2.58}$$

The kinetic energy of the system is

$$\begin{aligned} T &= \frac{1}{2}m\dot{s}^2 + \frac{1}{2}m(\dot{x}^2 + \dot{y}^2) \\ &= \frac{1}{2}m(\dot{s}^2 + (\dot{s} + d\cos\theta\,\dot{\theta})^2 + d^2\sin^2\theta\,\dot{\theta}^2) \\ &= m\dot{s}^2 + \frac{1}{2}md^2\dot{\theta}^2 + md\cos\theta\dot{s}\dot{\theta}\,, \end{aligned} \tag{2.59}$$

and the potential energy

$$V = mgy = -mgd\cos\theta. \tag{2.60}$$

This determines the Lagrangian $L = T - V$,

$$L = m\dot{s}^2 + \frac{1}{2}md^2\dot{\theta}^2 + md\cos\theta\dot{s}\dot{\theta} + mgd\cos\theta\,. \tag{2.61}$$

The partial derivatives of $L$ with respect to $\dot{\theta}$ and $\theta$ are

$$\begin{aligned} \frac{\partial L}{\partial \dot{\theta}} &= md^2\dot{\theta} + md\cos\theta\dot{s}\,, \\ \frac{\partial L}{\partial \theta} &= -md\sin\theta\dot{s}\dot{\theta} - mgd\sin\theta\,, \end{aligned} \tag{2.62}$$

which gives Lagrange's equation as

$$\frac{d}{dt}\frac{\partial L}{\partial \dot{\theta}} - \frac{\partial L}{\partial \theta} = 0 \quad \Rightarrow \quad \ddot{\theta} + \frac{1}{d}\cos\theta\ddot{s} + \frac{g}{d}\sin\theta = 0\,. \tag{2.63}$$

b) Since $s$ is cyclic, which means that $\frac{\partial L}{\partial s} = 0$, Lagrange's equation implies that the conjugate momentum is a constant of motion

$$p_s \equiv \frac{\partial L}{\partial \dot{s}} = 2m\dot{s} + md\cos\theta\dot{\theta}\,. \tag{2.64}$$

This implies

$$\ddot{s} = -\frac{d}{2}\frac{d}{dt}(\cos\theta\,\dot{\theta}) = -\frac{1}{2}d\cos\theta\,\ddot{\theta} + \frac{1}{2}d\sin\theta\,\dot{\theta}^2\,, \tag{2.65}$$

which can be used to eliminate the variable $s$ from the $\theta$ equation. The result is

$$(1 - \frac{1}{2}\cos^2\theta)\ddot{\theta} + \frac{1}{2}\sin\theta\cos\theta\,\dot{\theta}^2 + \frac{g}{d}\sin\theta = 0 \tag{2.66}$$

and we note, in particular, that the equation is independent of the value of the constant $p_s$.

c) A small angle approximation of this equation around the equilibrium point $\theta = 0$ means to expand all the terms of the equation to first order in $\theta$ and its time derivatives. We first note that the middle term in (2.66) gives no contribution to first order due to its quadratic dependence of $\dot{\theta}$. Using the first order approximations $\cos\theta \approx 1$ and $\sin\theta \approx \theta$, we get the following simplified equation

$$\ddot{\theta} + 2\frac{g}{d}\theta = 0\,. \tag{2.67}$$

This has the form of a harmonic oscillator equation with angular momentum $\omega = \sqrt{2g/d}$.

### *Problem 2.9*

A small body with mass $m$ is constrained to move along a spiral-shaped channel on a circular disk with radius $R$. The disk rotates in the horizontal plane, with constant angular velocity $\omega$ about an axis through the center of the disk. The points on the spiral are characterized by polar coordinates $(r, \theta)$, with $r = a\theta$, where $a$ is a constant, and $\theta$ is measured relative to a reference frame which rotates with the disk.

a) The Cartesian coordinates $(x, y)$ of a point on the spiral, measured in a non-rotating reference frame, are related to the coordinates of the rotating frame by

$$\begin{aligned} x &= r\cos(\theta + \omega t) = a\theta\cos(\theta + \omega t)\,, \\ y &= r\sin(\theta + \omega t) = a\theta\sin(\theta + \omega t)\,. \end{aligned} \tag{2.68}$$

The corresponding velocity components are

$$\begin{aligned} \dot{x} &= a\dot{\theta}\cos(\theta + \omega t) - a\theta(\dot{\theta} + \omega)\sin(\theta + \omega t)\,, \\ \dot{y} &= a\dot{\theta}\sin(\theta + \omega t) + a\theta(\dot{\theta} + \omega)\cos(\theta + \omega t)\,. \end{aligned} \tag{2.69}$$

The Lagrangian, which is identical to the kinetic energy, then is

$$\begin{aligned} L &= \frac{1}{2}m(\dot{x}^2 + \dot{y}^2) \\ &= \frac{1}{2}ma^2((1+\theta^2)\dot{\theta}^2 + 2\omega\theta^2\dot{\theta} + \omega^2\theta^2)\,. \end{aligned} \tag{2.70}$$

b) We find the partial derivatives

$$\begin{aligned} \frac{\partial L}{\partial \dot{\theta}} &= ma^2((1+\theta^2)\dot{\theta} + \omega\theta^2)\,, \\ \frac{\partial L}{\partial \theta} &= ma^2(\theta\dot{\theta}^2 + 2\omega\theta\dot{\theta} + \omega^2\theta)\,, \end{aligned} \tag{2.71}$$

and from these Lagrange's equation

$$\frac{d}{dt}\frac{\partial L}{\partial \dot{\theta}} - \frac{\partial L}{\partial \theta} = 0 \quad \Rightarrow$$

$$(1+\theta^2)\ddot{\theta} + \theta\dot{\theta}^2 - \omega^2\theta = 0\,. \tag{2.72}$$

For large angles, $\theta >> 1$, the following approximation is justified, $(1+\theta^2)\ddot{\theta} \approx \theta^2\ddot{\theta}$. This changes the equation of motion to

$$\theta\ddot{\theta} + \dot{\theta}^2 - \omega^2 = 0\,. \tag{2.73}$$

c) The simplified equation can be written as

$$\frac{1}{2}\frac{d^2}{dt^2}\theta^2 = \omega^2\,, \tag{2.74}$$

which has the solution

$$\theta^2 = \omega^2 t^2 + At + B\,, \tag{2.75}$$

with $A$ and $B$ as integration constants. With the initial conditions $r(0) = r_0$ and $\dot{r}(0) = 0$, which implies $\theta(0) = r_0/a$ and $\dot{\theta}(0) = 0$, we get $A = 0$ and $B = r_0^2/a^2$, the solution is

$$\theta(t) = \sqrt{\omega^2 t^2 + r_0^2/a^2}\,. \tag{2.76}$$

With $r_0 >> a$ this gives $\theta(t) >> 1$, which shows that the approximation which leads to Eq. (2.73) is satisfied.

d) The time needed for the body to reach the edge of the disk, is determined by $r(t) = R$, which means $\theta(t) = R/a$. This gives

$$R = \sqrt{a^2\omega^2 t^2 + r_0^2}\,, \tag{2.77}$$

which determines the time as

$$t = \frac{1}{a\omega}\sqrt{R^2 - r_0^2}\,. \tag{2.78}$$

***Problem 2.10***

We study here the motion of a particle with mass $m$ and charge $q$ in an electromagnetic field. The field is described by a time dependent vector potential $\mathbf{A}$, which in cylindrical coordinates has the form

$$A_r = A_z = 0,\ A_\phi = \frac{1}{2} r B(t). \tag{2.79}$$

a) The corresponding magnetic and electric fields, expressed in cylindrical coordinates, are

$$\begin{aligned} \mathbf{B} = \boldsymbol{\nabla}\times\mathbf{A} \quad &\Rightarrow \quad B_r = B_\phi = 0\,,\ B_z = \frac{1}{r}\frac{\partial}{\partial r}(rA_\phi) = B, \\ \mathbf{E} = -\frac{\partial}{\partial t}\mathbf{A} \quad &\Rightarrow \quad E_r = E_z = 0\,,\ E_\phi = -\frac{1}{2} r\frac{d}{dt}B\,. \end{aligned} \tag{2.80}$$

b) The Lagrangian of the charged particle is

$$L = \frac{1}{2}m\dot{\mathbf{r}}^2 + q\mathbf{A}\cdot\dot{\mathbf{r}} = \frac{1}{2}m(\dot r^2 + r^2\dot\phi^2) + \frac{1}{2}qBr^2\dot\phi\,, \tag{2.81}$$

with partial derivatives

$$\frac{\partial L}{\partial \dot r} = m\dot r\,, \quad \frac{\partial L}{\partial r} = mr\dot\phi^2 + qBr\dot\phi\,, \tag{2.82}$$

and

$$\frac{\partial L}{\partial \dot\phi} = mr^2\dot\phi + \frac{1}{2}qBr^2\,, \quad \frac{\partial L}{\partial \phi} = 0\,. \tag{2.83}$$

Lagrange's equation gives the following equation of motion for the $r$ variable,

$$\frac{d}{dt}\left(\frac{\partial L}{\partial \dot r}\right) - \frac{\partial L}{\partial r} = 0 \quad \Rightarrow \quad \ddot r - r\dot\phi^2 - \frac{qB}{m}r\dot\phi = 0\,, \tag{2.84}$$

and for the $\phi$ variable,

$$\frac{d}{dt}\left(\frac{\partial L}{\partial \dot\phi}\right) - \frac{\partial L}{\partial \phi} = 0 \quad \Rightarrow$$

$$r\ddot\phi + 2\dot r(\dot\phi + \frac{1}{2}\frac{qB}{m}) + \frac{1}{2}\frac{q}{m}r\frac{dB}{dt} = 0\,. \tag{2.85}$$

The standard expression for the equation of motion on vector form is

$$m\ddot{\mathbf{r}} = q(\mathbf{E} + \dot{\mathbf{r}}\times\mathbf{B})\,. \tag{2.86}$$

Decomposition of both sides of the equation in radial and angular parts gives

$$\ddot{\mathbf{r}} = (\ddot r - r\dot\phi^2)\mathbf{e}_r + (r\ddot\phi + 2\dot r\dot\phi)\mathbf{e}_\phi\,,$$

$$\frac{q}{m}(\mathbf{E} + \dot{\mathbf{r}}\times\mathbf{B}) = r\dot\phi\frac{qB}{m}\mathbf{e}_r - (\dot r\frac{qB}{m} + \frac{1}{2}r\frac{q}{m}\frac{dB}{dt})\mathbf{e}_\phi\,, \tag{2.87}$$

and equality of the radial and angular components separately reproduces the Lagrange's equations in (2.84) and (2.85).

c) With $B$ constant there are two constants of motion:
1) Since $\phi$ is cyclic the conjugate momentum is a constant,

$$\frac{\partial L}{\partial \phi} = 0 \quad \Rightarrow \quad \frac{\partial L}{\partial \dot{\phi}} = mr^2\dot{\phi} + \frac{1}{2}qBr^2 \equiv \ell\,(\text{const})\,. \tag{2.88}$$

2) $B = B_0$ implies that $L$ is time independent, and the Hamiltonian $H$ is therefore a constant of motion,

$$\begin{aligned} \frac{\partial L}{\partial t} &= 0 \quad \Rightarrow \\ H &= \frac{\partial L}{\partial \dot{r}}\dot{r} + \frac{\partial L}{\partial \dot{\phi}}\dot{\phi} - L \\ &= \frac{1}{2}m(\dot{r}^2 + r^2\dot{\phi}^2) \equiv \mathcal{E}\,(\text{const})\,. \end{aligned} \tag{2.89}$$

We assume circular motion, $r = r_0$. Inserted in the radial equation this gives

$$\ddot{r} - r\dot{\phi}^2 - \frac{qB}{m}r\dot{\phi} = -r_0\dot{\phi}(\dot{\phi} + \frac{qB_0}{m}) = 0\,, \tag{2.90}$$

and in the angular equation

$$r\ddot{\phi} + 2\dot{r}(\dot{\phi} + \frac{1}{2}\frac{qB}{m}) + \frac{1}{2}\frac{q}{m}r\frac{dB}{dt} = r_0\ddot{\phi} = 0\,. \tag{2.91}$$

Both equations are satisfied, provided the angular frequency is constant with value

$$\dot{\phi} = -\frac{qB_0}{m} = \omega_0\,. \tag{2.92}$$

The two constants of motion in this case take the values

$$\ell = \frac{1}{2}mr_0^2\omega_0\,, \quad \mathcal{E} = \frac{1}{2}mr_0^2\omega_0^2\,. \tag{2.93}$$

d) We assume now that $B$ is slowly changing with time, from an initial value $B_0$ to a final value $B_1$. The circular motion will then change from the initial radius $r_0$ to the final radius $r_1$. When $B$ changes $H$ is no longer constant, while $\ell$ continues to be constant since $\phi$ remains cyclic. This can be used to relate $r_0$ to $r_1$,

$$r_0^2\omega_0 = r_1^2\omega_1 \quad \Rightarrow \quad \frac{r_1}{r_0} = \sqrt{\frac{\omega_0}{\omega_1}} = \sqrt{\frac{B_0}{B_1}}\,. \tag{2.94}$$

For the initial and final energies this gives

$$\frac{\mathcal{E}_1}{\mathcal{E}_0} = \frac{r_1^2\omega_1^2}{r_0^2\omega_0^2} = \frac{B_1}{B_0}\,. \tag{2.95}$$

***Problem 2.11***

A particle moves on a parabolic surface given by the equation $z = (\lambda/2)(x^2+y^2)$, where $z$ is the Cartesian coordinate in the vertical direction and $\lambda$ is a constant. The particle has mass $m$ and moves without friction on the surface under influence of gravitation. The particle's position is given by the polar coordinates $(r,\theta)$ of the *projection* of the position vector into the $x, y$ plane.

a) The coordinates $x$ and $y$ of the plane, when expressed in polar coordinates, give

$$x = r\cos\theta\,,\; y = r\cos\theta \quad \Rightarrow \quad \dot{x}^2+\dot{y}^2 = \dot{r}^2 + r^2\dot{\theta}^2\,. \tag{2.96}$$

The vertical coordinate $z$, when restricted to the parabolic surface, depends on $r$ and $\theta$ as

$$z = \frac{1}{2}\lambda(x^2+y^2) = \frac{1}{2}\lambda r^2 \quad \Rightarrow \quad \dot{z} = \lambda r\dot{r}\,. \tag{2.97}$$

This gives for the Lagrangian,

$$\begin{aligned} L &= \frac{1}{2}m(\dot{x}^2+\dot{y}^2+\dot{z}^2) - mgz \\ &= \frac{1}{2}m[(1+\lambda^2r^2)\dot{r}^2 + r^2\dot{\theta}^2 - g\lambda r^2] \end{aligned} \tag{2.98}$$

and the partial derivatives of this with respect to $r$ and $\dot{r}$ are

$$\begin{aligned} \frac{\partial L}{\partial r} &= m(\lambda^2 r\dot{r}^2 + r\dot{r} - g\lambda r)\,, \\ \frac{\partial L}{\partial \dot{r}} &= m(1+\lambda^2r^2)\dot{r}\,. \end{aligned} \tag{2.99}$$

Lagrange's equation for the radial variable is then

$$\begin{gathered} \frac{d}{dt}\frac{\partial L}{\partial \dot{r}} - \frac{\partial L}{\partial r} = 0 \quad \Rightarrow \\ (1+\lambda^2r^2)\ddot{r} + \lambda^2 r\dot{r}^2 - r\dot{\theta}^2 + g\lambda r = 0\,. \end{gathered} \tag{2.100}$$

Since the $\theta$ variable is cyclic, we have

$$\frac{\partial L}{\partial \theta} = 0 \quad \Rightarrow \quad \frac{d}{dt}\frac{\partial L}{\partial \dot{\theta}} = \frac{d}{dt}(mr^2\dot{\theta}) = 0\,, \tag{2.101}$$

which can be integrated to give

$$\dot{\theta} = \frac{\alpha}{r^2}\,, \tag{2.102}$$

with $\alpha$ as an integration constant.

b) This result can be used to eliminate the $\theta$ variable from the radial equation, which then takes the form

$$(1+\lambda^2 r^2)\ddot{r} + \lambda^2 r \dot{r}^2 - \frac{\alpha^2}{r^3} + g\lambda r = 0\,. \tag{2.103}$$

This has a circular solution, $r = r_0, \dot{r} = \ddot{r} = 0$, provided the following equation is satisfied,

$$\frac{\alpha^2}{r_0^4} = g\lambda \quad \Rightarrow \quad \dot{\theta} = \sqrt{g\lambda}\,. \tag{2.104}$$

c) We consider small deviations from the circular motion by setting $r = r_0 + \rho, \dot{r} = \dot{\rho}, \ddot{r} = \ddot{\rho}$, and keeping only linear terms in $\rho$ and its time derivatives in the radial equation. This gives

$$\begin{aligned} &(1+\lambda^2 r_0^2)\ddot{\rho} + (3\frac{\alpha^2}{r_0^4} + g\lambda)\rho = 0 \\ \Rightarrow \quad &\ddot{\rho} + 4\frac{g\lambda}{1+\lambda^2 r_0^2}\rho = 0\,. \end{aligned} \tag{2.105}$$

This is a harmonic oscillator equation with angular frequency

$$\omega = 2\sqrt{\frac{g\lambda}{1+\lambda^2 r_0^2}}\,. \tag{2.106}$$

The solution to the harmonic oscillator equation can be written as

$$\rho(t) = \rho_0 \cos[\omega(t-t_0)] \tag{2.107}$$

which gives

$$z(t) = \frac{1}{2}\lambda r_0^2 + \lambda r_0 \rho_0 \cos[\omega(t-t_0)]\,, \tag{2.108}$$

and

$$\dot{\theta}(t) = \sqrt{g\lambda}(1 - 2\frac{\rho_0}{r_0}\cos[\omega(t-t_0)])\,. \tag{2.109}$$

This describes motion with small oscillations in the vertical coordinate $z$ around the value $\frac{1}{2}\lambda r_0^2$ combined with small oscillations in the angular velocity around the value $\sqrt{gl}$.

## *Problem 2.12*

A small body with mass $m$ and charge $q$ is moving in the horizontal plane ($x, y$-plane), under influence of a harmonic oscillator potential, $V(r) = \frac{1}{2}m\omega_0^2 r^2$ and a constant magnetic field $\mathbf{B} = B\,\mathbf{k}$, which is directed perpendicular to the plane of the moving particle. The vector potential corresponding to $\mathbf{B}$ can be written as $\mathbf{A} = -\frac{1}{2}\mathbf{r}\times\mathbf{B}$, with $\mathbf{r}$ as the position vector of the particle.

a) We treat the small body as point-like. The Lagrangian is

$$\begin{aligned} L &= \frac{1}{2}m\mathbf{v}^2 - V(\mathbf{r}) + q\mathbf{v}\cdot\mathbf{A} \\ &= \frac{1}{2}m\dot{\mathbf{r}}^2 - \frac{1}{2}m\omega_0^2 r^2 - \frac{1}{2}q\mathbf{v}\cdot(\mathbf{r}\times\mathbf{B}) \\ &= \frac{1}{2}m(\dot{r}^2 + r^2\dot{\phi}^2) - \frac{1}{2}m\omega_0^2 r^2 + \frac{1}{2}qBr^2\dot{\phi} \\ &= \frac{1}{2}m(\dot{r}^2 + r^2(\dot{\phi}^2 - \omega_B\dot{\phi} - \omega_0^2)), \end{aligned} \tag{2.110}$$

with $\omega_B = -qB/m$.

b) The polar angle $\phi$ is cyclic, which means that it does not appear in the Lagrangian. The conjugate momentum $p_\phi$ is then a constant of motion,

$$p_\phi = \frac{\partial L}{\partial\dot{\phi}} = mr^2(\dot{\phi} - \frac{1}{2}\omega_B) \equiv \ell\,. \tag{2.111}$$

$\ell$ can be interpreted as the conserved angular momentum. It has two contributions, the mechanical angular momentum, which is proportional to $\dot{\phi}$, and an electromagnetic field contribution, which is proportional to $qB$.

Since the Lagrangian is time independent the Hamiltonian is a conserved quantity. Here, it has the form

$$H = p_r\dot{r} + p_\phi\dot{\phi} - L = \frac{1}{2}m(\dot{r}^2 + r^2(\dot{\phi}^2 + \omega_0^2)) = T + V\,, \tag{2.112}$$

and is interpreted as the conserved energy of the system.

c) Lagrange's equation for the variable $r$ is

$$\frac{d}{dt}\frac{\partial L}{\partial\dot{r}} - \frac{\partial L}{\partial r} = 0 \quad\Rightarrow\quad \ddot{r} - r(\dot{\phi}^2 - \omega_B\dot{\phi} - \omega_0^2) = 0\,. \tag{2.113}$$

We note that the squared angular momentum is

$$\ell^2 = m^2 r^4(\dot{\phi}^2 - \omega_B\dot{\phi} + \frac{1}{4}\omega_B^2)\,. \tag{2.114}$$

This expression can be used to remove $\dot{\phi}$ from the radial equation, which then takes the form

$$\ddot{r} - \frac{\ell^2}{m^2 r^3} + r(\omega_0^2 + \frac{1}{4}\omega_B^2) = 0\,. \tag{2.115}$$

d) When $\ell \neq 0$ the radial equation has solutions corresponding to circular motion in the plane, with radius $r_0$ given by

$$r_0 = \sqrt{\frac{\ell}{m\Omega}}\,, \quad \Omega = \sqrt{\omega_0^2 + \frac{1}{4}\omega_B^2}\,. \tag{2.116}$$

The corresponding expression for $\dot{\phi}$, as shown by (2.111), is

$$\dot{\phi} = \frac{\ell}{mr_0^2} + \frac{1}{2}\omega_B = \Omega + \frac{1}{2}\omega_B\,. \tag{2.117}$$

When $\ell = 0$ the radial equation (2.115) is reduced to

$$\ddot{r} + \Omega^2 r = 0\,, \tag{2.118}$$

with solutions of the form $r = R\cos\Omega t$, and with angular velocity $\dot{\phi} = \frac{1}{2}\omega_B$. It shows harmonic oscillations in the radial coordinate combined with a constant angular velocity of the particle. Note that radial frequency is higher than the angular frequency, since $\Omega \geq \frac{1}{2}|\omega_B|$.

When $\ell \neq 0$ we have $\ddot{r} < 0$ for $r > r_0$ and $\ddot{r} > 0$ for $r < 0$. This shows that the general solution describes oscillations in the radial coordinate, about $r_0 = \sqrt{\ell/m\Omega}$. Similarly, the angular velocity $\dot{\phi}$ will oscillate around the value $\Omega + \frac{1}{2}\omega_B$. This means that the general solution will be a periodic modulation of the special solution where the particle moves with constant angular velocity in a circular orbit.

### *Problem 2.13*

We study here the motion of a Foucault pendulum. The pendulum is situated at the latitude 60° north, and we have the following information about the pendulum: The length of the pendulum wire is $l = 14m$ and the mass of the brass sphere at the end of the wire is $m = 20kg$. The idea is to use Lagrange's formalism to study the effect of the earth's rotation on the motion of the pendulum.

A set of earth-fixed orthogonal unit vectors are introduced, $\mathbf{e}_k, k = 1, 2, 3$, with $\mathbf{e}_3$ pointing in the vertical direction, $\mathbf{e}_1$ pointing to the north, and $\mathbf{e}_2$ orthogonal to the two. The three unit vectors are used as the basis vectors of an earth-fixed reference frame $S$, with the origin of the reference frame taken as the equilibrium position of the pendulum sphere. In addition

$\mathbf{k}$ is a unit vector in the plane spanned by $\mathbf{e}_1$ and $\mathbf{e}_3$, with direction parallel to the earth's rotational axis. The angle between $\mathbf{e}_3$ and $\mathbf{k}$ is referred to as $\theta$.

a) The position and velocity vectors of the pendulum sphere are

$$\begin{aligned}\mathbf{r} &= x\mathbf{e}_1 + y\mathbf{e}_2 + z\mathbf{e}_3\,,\\ \dot{\mathbf{r}} &= \dot{x}\mathbf{e}_1 + \dot{y}\mathbf{e}_2 + \dot{z}\mathbf{e}_3 + x\boldsymbol{\omega}\times\mathbf{e}_1 + y\boldsymbol{\omega}\times\mathbf{e}_2 + z\boldsymbol{\omega}\times\mathbf{e}_3\,,\end{aligned} \tag{2.119}$$

with

$$\boldsymbol{\omega} = \omega\mathbf{k}\,,\quad \mathbf{k} = \sin\theta\mathbf{e}_1 + \cos\theta\mathbf{e}_3\,, \tag{2.120}$$

where $\omega$ is the angular velocity of the rotating earth. This gives

$$\begin{aligned}\mathbf{k}\times\mathbf{e}_1 &= \cos\theta\,\mathbf{e}_2\,,\\ \mathbf{k}\times\mathbf{e}_2 &= -\cos\theta\,\mathbf{e}_1 + \sin\theta\mathbf{e}_3\,,\\ \mathbf{k}\times\mathbf{e}_3 &= -\sin\theta\,\mathbf{e}_2\,,\end{aligned} \tag{2.121}$$

and

$$\dot{\mathbf{r}} = (\dot{x} - \omega y\cos\theta)\mathbf{e}_1 + (\dot{y} + \omega(x\cos\theta - z\sin\theta))\mathbf{e}_2 + (\dot{z} + \omega y\sin\theta)\mathbf{e}_3\,. \tag{2.122}$$

The kinetic energy is

$$\begin{aligned}T = \frac{1}{2}m\,\big[&\dot{x}^2 + \dot{y}^2 + \dot{z}^2 + 2\omega\cos\theta(x\dot{y} - y\dot{x})\\ &+2\omega\sin\theta(y\dot{z} - z\dot{y}) + \mathcal{O}(\omega^2)\big]\,.\end{aligned} \tag{2.123}$$

To judge the importance of higher order terms in $\omega$, the relevant comparison is with the angular frequency of the pendulum, $\omega_p$. The periods of these two are

$$T_{earth} = 1\,\text{day} = 8640\,\text{s}\,,\quad T_p = \sqrt{g/l} = 0.84\,\text{s}\,. \tag{2.124}$$

This gives

$$\frac{\omega}{\omega_p} = \frac{T_p}{T_{earth}} \approx 10^{-4} \tag{2.125}$$

which shows that the quadratic terms in $\omega$ can safely be neglected. The potential energy of the pendulum is

$$V = mgz. \tag{2.126}$$

b) The $z$ coordinate can be expressed in terms of $x$ and $y$ in the following way,

$$\begin{aligned}&z^2 - 2lz + x^2 + y^2 = 0 \quad\Rightarrow\\ &z = l - \sqrt{l^2 - x^2 - y^2} \approx \frac{1}{2l}(x^2 + y^2)\,,\end{aligned} \tag{2.127}$$

where we have applied the small oscillation approximation. This shows that $z/l$ is second order in the small quantities $x/l$ and $y/l$. Therefore it is sufficient, in the Lagrangian, to include only first order terms in $z$, which gives

$$L = \frac{1}{2}m(\dot{x}^2 + \dot{y}^2) + 2\omega \cos\theta(x\dot{y} - y\dot{x}) - \frac{1}{2l}mg(x^2 + y^2) \,. \quad (2.128)$$

c) We change to polar coordinates,

$$\begin{aligned} x &= \rho\cos\phi\,, \quad \dot{x} = \dot{\rho}\cos\phi - \rho\dot{\phi}\sin\phi\,, \\ y &= \rho\sin\phi\,, \quad \dot{y} = \dot{\rho}\sin\phi + \rho\dot{\phi}\cos\phi\,. \end{aligned} \quad (2.129)$$

This gives the following expression for the Lagrangian

$$L = \frac{1}{2}(\dot{\rho}^2 + \rho^2\dot{\phi}^2) + m\omega\cos\theta\rho^2\dot{\phi} - \frac{1}{2l}mg\rho^2 \,. \quad (2.130)$$

Since the angle $\phi$ is cyclic, the corresponding conjugate momentum is a constant of motion,

$$\frac{\partial L}{\partial \dot{\phi}} = m\rho^2\dot{\phi} + m\rho^2\omega\cos\theta \equiv k \,. \quad (2.131)$$

Solved for the angular velocity this gives

$$\dot{\phi} = -\omega\cos\theta + \frac{k}{m\rho^2} \,. \quad (2.132)$$

Lagrange's equation for the radial variable is

$$\begin{aligned} &\frac{d}{dt}\frac{\partial L}{\partial \dot{\rho}} - \frac{\partial L}{\partial \rho} = 0 \quad \Rightarrow \\ &m\ddot{\rho} - m\rho\dot{\phi}^2 - 2m\omega\cos\theta\rho\dot{\phi} + m\frac{g}{l}\rho = 0 \,. \end{aligned} \quad (2.133)$$

d) We make now the assumption that the angular velocity $\omega_\phi = \dot{\phi}$ is constant and that $\rho$ oscillates with time. As shown by (2.132) this happens only if $k = 0$, and it determines the value of the angular velocity as

$$\omega_\phi = -\omega\cos\theta \,. \quad (2.134)$$

The radial equation is then reduced to

$$\ddot{\rho} + \rho(\frac{g}{l} + \omega^2\cos^2\theta) = 0 \,. \quad (2.135)$$

This is a harmonic oscillator equation for the pendulum, with angular frequency

$$\omega_p = \sqrt{\frac{g}{l} + \omega^2\cos^2\theta} = \sqrt{\frac{g}{l}} + \mathcal{O}(\omega^2) \,. \quad (2.136)$$

e) The rotation of the plane of oscillations of the pendulum, relative to the building, is determined by the angle $\phi$. In one period of the rotation of the earth the rotation angle of the pendulum plane is

$$\Delta\phi = \omega_\phi T_{earth} = -\omega T_{earth} \cos\theta = -2\pi\cos\theta\,. \tag{2.137}$$

Expressed in radians, and with $\theta = 30°$, corresponding to the latitude $60°$ for the position of the pendulum, we get

$$\Delta\phi = -360° \cos(30°) = -312°\,. \tag{2.138}$$

# Chapter 3

# Hamiltonian dynamics

*Problem 3.1*

In Problem 2.12 the following Lagrangian has been introduced

$$L = \frac{1}{2}m(\dot{r}^2 + r^2(\dot{\phi}^2 - \omega_B\dot{\phi} - \omega_0^2)). \tag{3.1}$$

It describes the motion of a charged particle in a combination of a harmonic oscillator potential and a constant magnetic field, with $\omega_0$ as the harmonic oscillator angular frequency and $\omega_B$ as the cyclotron angular frequency. We will here study the Hamiltonian description of the same system.

a) To find the Hamiltonian corresponding to the Lagrangian (3.1) we need the canonical momenta corresponding to the variables $r$ and $\phi$,

$$p_r = \frac{\partial L}{\partial \dot{r}} = m\dot{r}\,, \quad p_\theta = \frac{\partial L}{\partial \dot{\theta}} = mr^2(\dot{\theta} - \frac{1}{2}\omega_B)\,. \tag{3.2}$$

The general definition of the Hamiltonian then gives

$$\begin{aligned}
H &= p_r\dot{r} + p_\theta\dot{\theta} - L \\
&= m\dot{r}^2 + mr^2(\dot{\theta}^2 - \frac{1}{2}\omega_B\dot{\theta}) - \frac{1}{2}m\dot{r}^2 - \frac{1}{2}mr^2(\dot{\theta}^2 - \omega_B\dot{\theta} - \omega_0^2) \\
&= \frac{1}{2}m\dot{r}^2 + \frac{1}{2}mr^2(\dot{\theta}^2 + \omega_0^2) \\
&= \frac{1}{2m}p_r^2 + \frac{1}{2}mr^2(\frac{p_\theta}{mr^2} + \frac{1}{2}\omega_B)^2 + \frac{1}{2}mr^2\omega_0^2 \\
&= \frac{1}{2m}p_r^2 + \frac{1}{2mr^2}p_\theta^2 + \frac{1}{2}\omega_B p_\theta + \frac{1}{2}mr^2(\omega_0^2 + \frac{1}{4}\omega_B^2)\,.
\end{aligned} \tag{3.3}$$

b) We derive from this Hamilton's equations of the system,

$$\begin{aligned}
\dot{r} &= \frac{\partial H}{\partial p_r} = \frac{p_r}{m}\,, \quad \dot{p}_r = -\frac{\partial H}{\partial r} = \frac{p_\theta^2}{mr^3} - mr(\omega_0^2 + \frac{1}{4}\omega_B^2) \\
\dot{\theta} &= \frac{\partial H}{\partial p_\theta} = \frac{p_\theta}{mr^2} + \frac{1}{2}\omega_B\,, \quad \dot{p}_\theta = -\frac{\partial H}{\partial \theta} = 0\,.
\end{aligned} \tag{3.4}$$

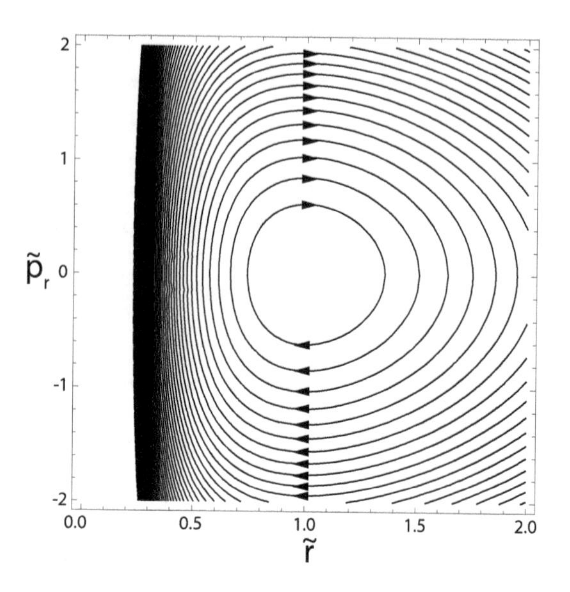


Fig. 3.1 Phase space plot, Problem 3.1

There are two constants of motion.

1) Angular momentum

$$\dot{p}_\theta = 0 \quad \Rightarrow \quad p_\theta = mr^2(\dot{\theta} - \frac{1}{2}\omega_B) \equiv \ell \,. \tag{3.5}$$

We note that in this expression, in addition to the standard term interpreted as the (mechanical) angular momentum of the particle, there is a term, proportional to $\omega_B$, which comes from the coupling of the electric charge to the magnetic field. This can be interpreted as an electromagnetic field contribution to the angular momentum.

2) Energy,

$$\frac{dH}{dt} = \frac{\partial H}{\partial t} = 0 \quad \Rightarrow$$
$$H = \frac{1}{2m}p_r^2 + \frac{\ell^2}{2mr^2} + \frac{1}{2}mr^2\Omega^2 - \frac{1}{2}\omega_B\ell \equiv E\,(\text{const})\,, \tag{3.6}$$

with $p_\theta$ replaced by $\ell$ and $\Omega = \sqrt{\omega_0^2 + \omega_B^2/4}$.

c) In dimensionless units, and with the constant term ($-\omega_B\ell/2$) omitted, the energy function can be written as

$$\tilde{E} = \frac{1}{2}\tilde{p}_r^2 + \frac{1}{2}(\frac{1}{\tilde{r}^2} + \tilde{r}^2), \tag{3.7}$$

with

$$\tilde{r} = \sqrt{\frac{m\Omega}{\ell}}\, r\,, \quad \tilde{p}_r = \frac{1}{\sqrt{m\ell\Omega}} p_r. \tag{3.8}$$

The expression (3.6) is used in the phase-space plot in Fig. 3.1. The relation, $\dot{r} = p_r/m$, implies that $\dot{r}$ is positive in the upper half-plane and negative in the lower half-plane. This means that the circulation around the equilibrium point has negative orientation, indicated by arrows in the plot.

This circulation in the (reduced) phase space corresponds to oscillations in the radial coordinate under the motion in the two-dimensional, physical plane. The center of the oscillations corresponds to $\tilde{r} = 1$ or $r = \sqrt{\ell/m\Omega}$ (see also Problem 2.12).

### *Problem 3.2*

We study here the motion of a particle with mass $m$, which moves in a one-dimension potential,

$$V(x) = \frac{1}{4}ax^4 - \frac{1}{2}kx^2\,, \tag{3.9}$$

with $a$ and $k$ as positive constants, and $x$ as the position coordinate of the particle.

a) The Lagrangian is

$$L = T - V = \frac{1}{2}m\dot{x}^2 - \frac{1}{4}ax^4 + \frac{1}{2}kx^2\,, \tag{3.10}$$

and the corresponding Lagrange's equation is

$$\frac{d}{dt}\frac{\partial L}{\partial \dot{x}} - \frac{\partial L}{\partial x} = 0 \quad \Rightarrow \quad m\ddot{x} + ax^3 - kx = 0\,. \tag{3.11}$$

b) The equilibrium points are points $x$, where the equation of motion is satisfied with $\ddot{x} = \dot{x} = 0$. The points are thus determined by the equation

$$ax^3 - kx = 0 \quad \Rightarrow \quad x = \{\pm\sqrt{\frac{k}{a}}, 0\}\,. \tag{3.12}$$

The point $x = 0$ is an unstable equilibrium since $\frac{d^2V}{dx^2}(0) = -k < 0$, and the points $x = \pm\sqrt{k/a} \equiv x_\pm$ are stable equilibria, since $\frac{d^2V}{dx^2}(x_\pm) = 2k > 0$.

Assume $x = x_\pm + \eta$, with $\eta$ as a small deviation from the equilibrium point. With $\eta$ included only to first order in the equation of motion, this gives

$$m\ddot{\eta} + 3ax_\pm^2\eta - k\eta = 0 \quad \Rightarrow \quad \ddot{\eta} + 2\frac{k}{m}\eta = 0\,. \tag{3.13}$$

This shows that the angular frequency for small oscillations about the stable equilibria is $\omega = \sqrt{2k/m}$.

c) The conjugate momentum is $p = \frac{\partial L}{\partial \dot{x}} = m\dot{x}$, and the Hamiltonian then is

$$H(x,p) = p\dot{x} - L = \frac{p^2}{2m} + \frac{1}{4}ax^4 - \frac{1}{2}kx^2 \,. \tag{3.14}$$

Hamilton's equations are

$$\dot{x} = \frac{\partial H}{\partial p} = \frac{p}{m}, \quad \dot{p} = -\frac{\partial H}{\partial x} = ax^3 - kx \,. \tag{3.15}$$

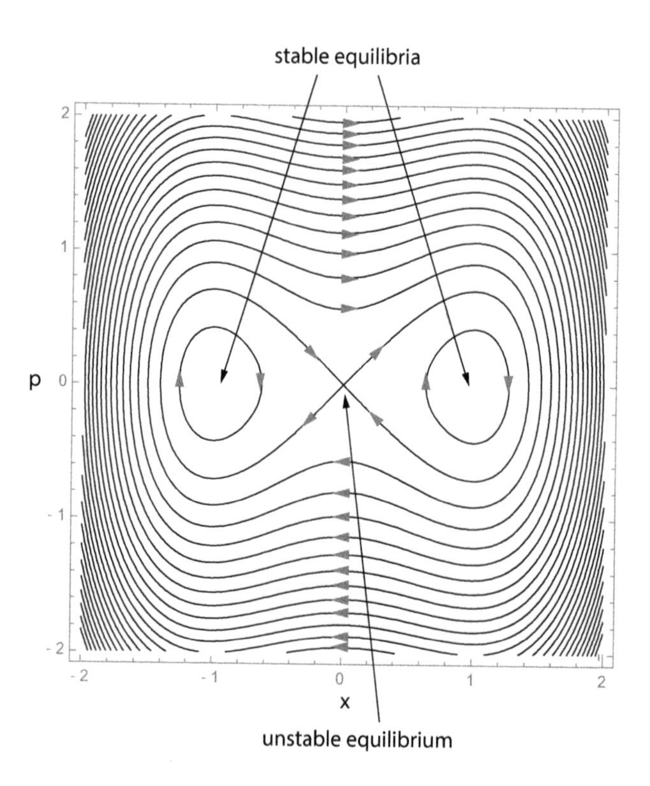


Fig. 3.2 Phase space plot of model with two stable equilibria. Dimensionless variables are used, with $m = a = k = 1$.

d) The plot in Fig. 3.2 shows the equipotential curves of the energy function $H(x,p)$, with the arrows pointing in the directions of the phase space flow. There are two different types of motion. For low energies the closed curves correspond to oscillations about one of the stable equilibrium points, while for higher energies both equilibrium points will be passed in the oscillations.

***Problem 3.3***

We consider here a particle of mass $m$, which moves in a one-dimensional periodic potential

$$V(x) = V_0(\sin x + a\sin^2 x)\,, \tag{3.16}$$

where $x$ is the coordinate of the particle in the direction of motion, $a > 0$ is an external parameter, which can be varied, and where $V_0$ is a constant which measures the strength of the potential.

a) The potential and its derivatives are

$$\begin{aligned} V(x) &= V_0(\sin x + a\sin^2 x),\\ V'(x) &= V_0(\cos x + a\sin 2x),\\ V''(x) &= V_0(-\sin x + 2a\cos 2x)\,. \end{aligned} \tag{3.17}$$

There are two types of equilibrium points, determined by $V'(x) = 0$,

$$\begin{aligned} &\text{I}\;\; \cos x = 0 \quad \Rightarrow \quad x = (n+\frac{1}{2})\pi\,, \quad n = 0, \pm 1, \dots\,,\\ &\text{II}\;\; \sin x = -\frac{1}{2a} \quad \Rightarrow \quad x = -\arcsin(\frac{1}{2a})\,. \end{aligned} \tag{3.18}$$

For $a < \frac{1}{2}$ there are only type I solutions, but for $a > \frac{1}{2}$ there are both types of solutions.

The values of $V''(x)$ at the equilibrium points are

$$\begin{aligned} &\text{Type I}\;\; V'' = V_0((-1)^{n+1} - 2a)\,,\\ &\text{Type II}\;\; V'' = V_0(2a - \frac{1}{2a})\,. \end{aligned} \tag{3.19}$$

Stable equilibria are determined by $V'' > 0$, and the expressions above show that this is satisfied for the Type I equilibria if $a < \frac{1}{2}$ and $n$ is odd, and for the Type II equilibria if $a > \frac{1}{2}$.

We sum up the results concerning the equilibrium points, with $n$ taking the values $0, \pm 1 ...$,

$$\begin{aligned} a < \frac{1}{2} \quad & \text{stable equilibria}: \quad x = (2n + \frac{3}{2})\pi\,, \\ & \text{unstable equilibria}: \quad x = (2n + \frac{1}{2})\pi\,, \\ a > \frac{1}{2} \quad & \text{stable equilibria}: \quad x = -\arcsin(\frac{1}{2a})\,, \\ & \text{unstable equilibria}: \quad x = (n + \frac{1}{2})\pi\,. \end{aligned} \tag{3.20}$$

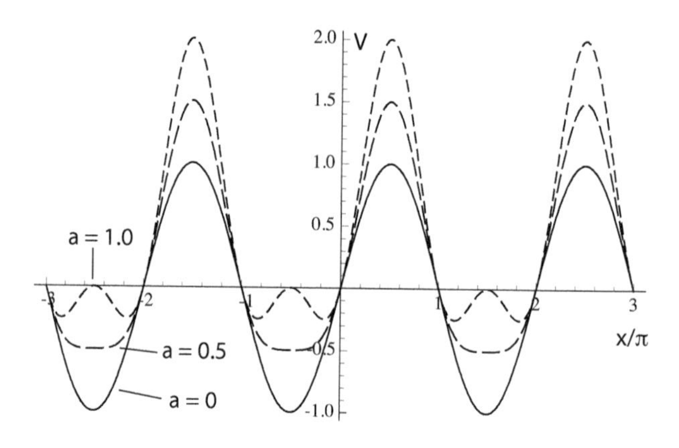


Fig. 3.3 The form of the potential in Problem 3.3, for three values of the parameter $a$. Dimensionless variables are used, with $m = V_0 = 1$.

b) The plot in Fig. 3.3 shows that when $a - 1/2$ changes from negative to positive, the stable equilibrium at $x = (2n + \frac{3}{2})\pi$ becomes unstable, and two new, stable equilibria appear in a symmetric way on both sides of these points.

c) The Lagrangian of the particle is

$$L = T - V = \frac{1}{2}m\dot{x}^2 - V_0(\sin x + a\sin^2 x)\,, \tag{3.21}$$

with Lagrange's equation

$$\frac{d}{dt}\frac{\partial L}{\partial \dot{x}} - \frac{\partial L}{\partial x} = 0 \quad \Rightarrow \quad m\ddot{x} + V'(x) = 0\,. \tag{3.22}$$

This gives

$$m\ddot{x} + V_0 \cos x(1 + 2a\sin x) = 0\,. \tag{3.23}$$

d) Let $x_0$ be any of the stable equilibrium points. To study small oscillations around the point, we write $x = x_0 + \xi$, and expand the potential to first order in $\xi$. This gives $V'(x) = V'(x_0) + V''(x_0)\xi + \ldots$, where $V'(x_0) = 0$ and $V''(x_0) > 0$. The equation of motion then takes the form

$$m\ddot{\xi} + V''(x_0)\xi = 0\,. \tag{3.24}$$

It has the form of a harmonic oscillator equation with angular frequency $\omega = \sqrt{V''(x_0)/m}$. The values of $V''(x_0)$ are determined in a),

$$\begin{aligned} a < \frac{1}{2}: \quad & V'' = V_0(1-2a) \quad \Rightarrow \quad \omega = \sqrt{\frac{V_0}{m}(1-2a)} \\ a > \frac{1}{2}: \quad & V'' = V_0(2a - \frac{1}{2a}) \quad \Rightarrow \quad \omega = \sqrt{\frac{V_0}{m}(2a - \frac{1}{2a})}\,. \end{aligned} \tag{3.25}$$

e) The Hamiltonian is $H = p\dot{x} - L$, with $p = \partial L/\partial \dot{x} = m\dot{x}$. This gives

$$H(x,p) = \frac{p^2}{2m} + V_0(\sin x + a\sin^2 x)\,. \tag{3.26}$$

$H(x,p)$ defines a phase-space potential, with the motion following equipotential curves. The direction and speed of the motion is determined by the gradient of the function,

$$(\dot{x}, \dot{p}) = (\partial H/\partial p, -\partial H/\partial x)\,. \tag{3.27}$$

f) A contour plot of the function $H(x,p)$ is shown in Fig. 3.4, for the three values $a = 0, 0.5$ and $1.0$. The dotted curves represent separatrices, which separate different types of motion. In the first case ($a = 0$) the diagram has the same form as for a planar pendulum, but here without the periodic identification of points along the $x$-axis. There are two types of motion, oscillations about the stable equilibria, and unbounded motion in the positive or negative $x$-direction for higher values of the energy.

The second case ($a = 0.5$) is a limit case where the energy minimum for $a < 0.5$ turns into a local maximum for $a > 0.5$. For $a = 0.5$ there still are only two types of motion, oscillations about the minimum for small energies and unbounded motion for higher energies. However, since in this case $V'' = 0$ at the potential minima, the small oscillations are not harmonic.

When $a = 1.0$ there are, as shown in the diagram, three types of motion. For sufficiently low energy, there will be small oscillations about a single equilibrium point. For somewhat higher energy there will be oscillations where the motion is bound to a pair of neighboring minima. Finally there are solutions with unbounded motion.

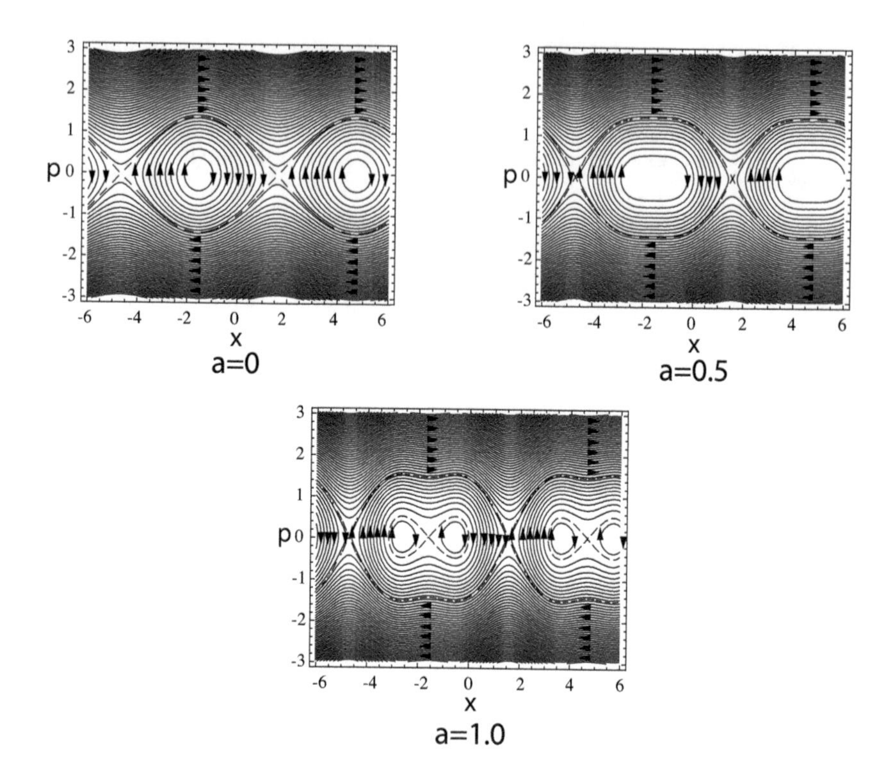


Fig. 3.4 Phase space plots of the model in Problem 3.3, for three values of the parameter $a$.

### *Problem 3.4*

In this problem we apply Fermat's principle to light rays restricted to a plane with Cartesian coordinates $(x, y)$. Fermat's principle determine the light paths as solutions to the variational problem, where the optical path length, defined (in the present case) as

$$S[y(x)] = \int_{x_1}^{x_2} n(x,y)\sqrt{1+y'^2}dx\ , \qquad y' = \frac{dy}{dx}, \tag{3.28}$$

is stationary. Here $n(x, y)$ is the position dependent refraction of the optical medium.

a) The variational problem can be solved by reformulating it as a Lagrangian problem. The Lagrangian in this problem is

$$L(y, y', x) = n(x,y)\sqrt{1+y'^2}\,. \tag{3.29}$$

The variational problem $\delta S = 0$, for variations in the path $y(x)$ with fixed end points $(x_1, y_1)$ and $(x_2, y_2)$, is equivalent to Lagrange's equation

$$\frac{d}{dx}\frac{\partial L}{\partial y'} - \frac{\partial L}{\partial y} = 0 \quad \Rightarrow$$

$$\frac{d}{dx}\left(\frac{n(x,y)y'}{\sqrt{1+y'^2}}\right) - \frac{\partial n}{\partial y}\sqrt{1+y'^2} = 0\,. \tag{3.30}$$

We first consider the case where $n$ is constant. The equation reduces to

$$\frac{d}{dx}\left(\frac{y'}{\sqrt{1+y'^2}}\right) = 0\,, \tag{3.31}$$

which implies that $y'' = 0$, and therefore that $y(x)$ is a straight line between the endpoints $(x_1, y_1)$ and $(x_2, y_2)$

b) We next make the assumption that the medium has two different, constant indices of refraction, with $n = n_1$ for $x < 0$ and $n = n_2$ for $x > 0$. This implies that the variational problem is a curve which is a straight line on both sides of the boundary $x = 0$ where $n$ changes value. The only variable to be determined by the variational problem is thus the crossing point $y_0 = y(0)$ at the boundary. The functional $S[y(x)]$ is then reduced to a function of $y_0$,

$$S(y_0) = n_1\sqrt{x_1^2 + (y_1 - y_0)^2} + n_2\sqrt{x_2^2 + (y_2 - y_0)^2}\,, \tag{3.32}$$

and the variational problem is reduced to finding a stationary point of this function,

$$\frac{dS}{dy_0} = 0 \quad \Rightarrow$$

$$n_1\frac{y_0 - y_1}{\sqrt{x_1^2 + (y_1 - y_0)^2}} + n_2\frac{y_0 - y_2}{\sqrt{x_1^2 + (y_1 - y_0)^2}} = 0\,. \tag{3.33}$$

c) With $\theta_1$ as the angle of the light ray for $x < 0$ measured relative to the normal of the boundary, and similarly $\theta_2$ as the angle of the light ray for $x > 0$, we have

$$\sin\theta_1 = \frac{y_1 - y_0}{\sqrt{x_1^2 + (y_1 - y_0)^2}}\,, \quad \sin\theta_2 = \frac{y_0 - y_2}{\sqrt{x_1^2 + (y_1 - y_0)^2}}\,, \tag{3.34}$$

and by use of these identities, Eq. (3.33) is simplified to

$$n_1 \sin\theta_1 = n_2 \sin\theta_2\,, \tag{3.35}$$

which we recognize as Snell's law of refraction.

## *Problem 3.5*

Fermat's principle is here applied to the light path in an optical medium, where the index of refraction $n(y)$ changes with the height $y$ in a vertical plane. With $x$ as the horizontal coordinate, the action integral is

$$S[y(x)] = \int_{x_1}^{x_2} n(y)\sqrt{1+y'^2}dx. \tag{3.36}$$

a) From the above expression follows that the corresponding Lagrangian can be identified as

$$L(y, y') = n(y)\sqrt{1+y'^2}\,. \tag{3.37}$$

We derive from this Lagrange's equation,

$$\begin{aligned} &\frac{d}{dx}\frac{\partial L}{\partial y'} - \frac{\partial L}{\partial y} = 0 \quad \Rightarrow \\ &\frac{d}{dx}\left(n(y)\frac{y'}{\sqrt{1+y'^2}}\right) - \frac{dn}{dy}\sqrt{1+y'^2} = 0\,. \end{aligned} \tag{3.38}$$

We perform the differentiations and simplify the expression in the following way,

$$\begin{aligned} &\frac{dn}{dy}\frac{y'^2}{\sqrt{1+y'^2}} + n(y)\frac{y''}{\sqrt{1+y'^2}} - n(y)\frac{y'^2y''}{(\sqrt{1+y'^2})^3} - \frac{dn}{dy}\sqrt{1+y'^2} = 0 \\ \Rightarrow \quad &-\frac{dn}{dy}\frac{1}{\sqrt{1+y'^2}} + n(y)\frac{y''}{(\sqrt{1+y'^2})^3} = 0\,. \end{aligned} \tag{3.39}$$

Solving this with respect to $y''$, we find

$$y'' = \frac{1}{n(y)}\frac{dn}{dy}(1+y'^2)\,. \tag{3.40}$$

b) We want to show that the following first order equation solves the second order equation (3.40)

$$\left(\frac{n(y)}{n_0}\right)^2 = 1 + y'^2\,, \tag{3.41}$$

with $n_0$ as a constant. To demonstrate this we differentiate the equation with respect to $x$. This gives

$$y''y' = \frac{n(y)}{n_0^2}\frac{dn}{dy}y'\,. \tag{3.42}$$

It has one (spurious) solution, $y' = 0$, which we disregard, and making use of (3.41) again, we reproduce, from (3.42), Eq. (3.40),

$$y'' = \left(\frac{n(y)}{n_0}\right)^2 \frac{1}{n(y)}\frac{dn}{dy}$$
$$= \frac{1}{n(y)}\frac{dn}{dy}(1 + y'^2). \tag{3.43}$$

c) The index of refraction is assumed to decrease with height inside the container, in the following way

$$n(y) = n_0\, e^{-\alpha y}. \tag{3.44}$$

At the point of entrance into the container the light beam is assumed to satisfy the conditions $y(0) = y'(0) = 0$, and $n(0) = n_0$. Inside the container, we have, as shown in b), the following relation

$$n(y) = \sqrt{1 + y'^2}\, n_0\,. \tag{3.45}$$

Since $y' = 0$ has been excluded as a false solution, we conclude from the expression above that $n(y) > n_0$ and increases along the path. This implies that the path is bent downwards when the light beam propagates through the solution.

d) We will check that the following expression gives a solution to Eq. (3.41) for the path of the light beam

$$e^{-\alpha y} = \frac{1}{\cos \alpha x}\,. \tag{3.46}$$

We first apply the relation to the left-hand side of (3.41)

$$\left(\frac{n(y)}{n_0}\right)^2 = e^{-2\alpha y} = \frac{1}{\cos^2 \alpha x}\,. \tag{3.47}$$

Next we find by differentiating (3.46) with respect to $x$,

$$-\,\alpha y' e^{-\alpha x} = \alpha \frac{\sin \alpha x}{\cos^2}\alpha x \quad \Rightarrow \quad y' = -\tan \alpha x\,. \tag{3.48}$$

When this is inserted in the right-hand side of (3.41) the result is

$$1 + y'^2 = 1 + \tan^2 \alpha x = \frac{1}{\cos^2 \alpha x}. \tag{3.49}$$

This confirms that (3.46) gives a solution to Eq. (3.41).

e) The deflection angle $\theta$ of the light beam is at the end of the container, $x = L$, given by

$$\tan\theta = \left(\frac{dy}{dx}\right)_{x=L} = -\tan \alpha L\,, \tag{3.50}$$

which gives $\theta = -\alpha L$, where the sign is consistent with the beam being deflected downwards.

### *Problem 3.6*

The brachistochrone problem is the following:
*Given two points A and B in a vertical plane, what is the curve traced out by a point acted on only by gravity, which starts at A and reaches B in the shortest time.*

We solve this as a variational problem, where the time $T$ spent on the path is written as

$$T[y(x)] = \int_{x_A}^{x_B} L(y, y')dx\,, \quad y' = \frac{dy}{dx}\,. \tag{3.51}$$

$y(x)$ is here the path followed by the particle, with $x$ as the horizontal coordinate, and $y$ as the vertical coordinate of the particle. The origin of the coordinate system is chosen with $x_A = y_A = 0$, and the potential energy is assumed to vanish at this point.

a) The first problem to be solved is to determine the form of the Lagrangian $L(y, y')$. We write the velocity of the particle as

$$v = \frac{ds}{dt} \quad \Rightarrow \quad dt = \frac{ds}{v}\,, \tag{3.52}$$

with

$$ds = \sqrt{dx^2 + dy^2} = \sqrt{1 + y'^2}dx\,, \quad y' \equiv \frac{dy}{dx}\,. \tag{3.53}$$

The velocity of the particle is assumed to vanish at the starting point, and energy conservation determines the velocity as a function of $y$,

$$\frac{1}{2}mv^2 + mgy = 0 \quad \Rightarrow \quad v = \sqrt{-2gy}\,, \tag{3.54}$$

where for convenience, we have chosen for the potential energy to vanish at $y = 0$. This determines the time spent on the path to be

$$\begin{aligned} T[y(x)] &= \int_{t_A}^{t_B} dt \\ &= \int_{x_A}^{x_B} \frac{1}{v}\frac{ds}{dx}dx \\ &= \int_{x_A}^{x_B} \sqrt{\frac{1 + y'^2}{-2gy}}dx\,, \end{aligned} \tag{3.55}$$

which gives the effective Lagrangian as

$$L(y, y') = \sqrt{\frac{1 + y'^2}{-2gy}}\,. \tag{3.56}$$

b) The coordinate $x$ in this problem has taken the place of time $t$ in the usual formulation of Lagrange's equations, and the momentum conjugate to $y$ is therefore

$$p = \frac{\partial L}{\partial y'} = \frac{y'}{(-2gy)(1+y'^2)}\,, \tag{3.57}$$

and the Hamiltonian is

$$\begin{aligned} H &= py' - L \\ &= \frac{y'^2}{(-2gy)(1+y'^2)} - \sqrt{\frac{1+y'^2}{-2gy}} \\ &= -\frac{1}{(-2gy)(1+y'^2)}\,. \end{aligned} \tag{3.58}$$

Since $H$ has no explicit $x$ dependence, it is a constant along the path in the $x,y$-plane. (This corresponds to energy conservation when $H$ is time independent in the standard Lagrange formulation.) This gives

$$(1+y'^2)y = -k^2\,, \tag{3.59}$$

with $k$ as a (positive) constant.

c) The assumption now is that paths defined by the parametric expressions

$$x = \frac{1}{2}k^2(\theta - \sin\theta)\,, \quad y = \frac{1}{2}k^2(\cos\theta - 1)\,, \tag{3.60}$$

are solutions of Eq. (3.59). To show that this is the case, we determine $y'$ from the above expressions,

$$y' = \frac{dy}{dx} = \frac{dy}{d\theta}\frac{d\theta}{dx} = \frac{dy}{d\theta}\Big/\frac{dx}{d\theta}\,. \tag{3.61}$$

The derivatives with respect to the parameter $\theta$ are

$$\begin{aligned} \frac{dx}{d\theta} &= \frac{1}{2}k^2(1-\cos\theta) = -y\,, \\ \frac{dy}{d\theta} &= -\frac{1}{2}k^2\sin\theta\,. \end{aligned} \tag{3.62}$$

This gives

$$\begin{aligned} (1+y'^2)y &= \left(1 + \frac{1}{y^2}\left(\frac{dy}{d\theta}\right)^2\right)y \\ &= \frac{1}{y}\left(y^2 + \left(\frac{dy}{d\theta}\right)^2\right) \\ &= \frac{k^4}{4y}((\cos\theta - 1)^2 + \sin^2\theta) \\ &= \frac{k^4}{2y}(1-\cos\theta) = -k^2\,. \end{aligned} \tag{3.63}$$

The boundary conditions at the starting point, $x_A = y_A = 0$, are satisfied if the start value of the path parameter is $\theta_A = 0$. This is clear from (3.60). The boundary conditions at the end point are

$$x_B = \frac{1}{2}k^2(\theta_B - \sin\theta_B)\,, \quad y_B = \frac{1}{2}k^2(\cos\theta_B - 1)\,. \tag{3.64}$$

These equations determine the two free parameters of the solution, $k^2$ and $\theta_B$.

d) The form of the path, as a cycloid, is shown in Fig. 3.5.

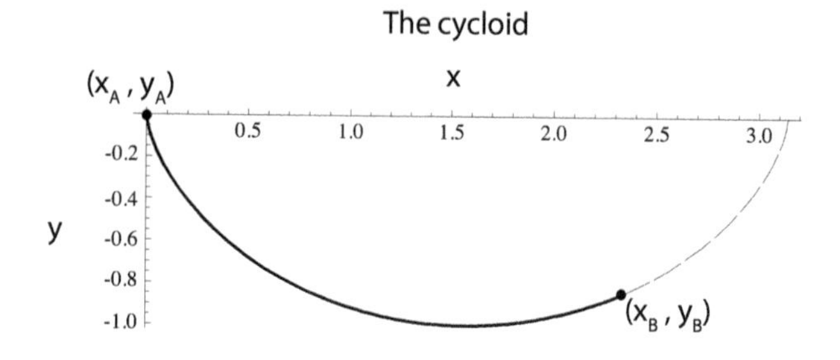


Fig. 3.5 The cycloid defines the path which minimizes the transit time between the chosen endpoints. The parameter $k$ is set to 1, and the end point $(x_B, y_B)$ is chosen arbitrarily on the curve.

e) At the bottom of the cycloid we have

$$\frac{dy}{d\theta} = 0 \quad \Rightarrow \quad \sin\theta = 0 \quad \Rightarrow \quad \theta_B = \pi\,. \tag{3.65}$$

With this as the end point of the path, we have

$$x_B = \frac{1}{2}\pi k^2\,, \quad y_B = -k^2 \quad \Rightarrow \quad y_B = -\frac{2}{\pi}x_B\,. \tag{3.66}$$

The time spent on the path is in this case

$$\begin{aligned} T &= \int_0^{x_B} \sqrt{\frac{1+y'^2}{-2gy}}dx = \int_0^{\pi} \sqrt{\frac{1+y'^2}{-2gy}}\frac{dx}{d\theta}d\theta \\ &= \int_0^{\pi} \sqrt{\frac{k^2}{2gy^2}}(-y)d\theta = \frac{k}{\sqrt{2g}}\int_0^{\pi} d\theta = \frac{\pi k}{\sqrt{2g}}\,. \end{aligned} \tag{3.67}$$

The length of the straight line between the end points of the path is

$$s = \sqrt{x_B^2 + y_B^2} = k^2\sqrt{\frac{\pi^2}{4} + 1}\,. \tag{3.68}$$

If the particle follows this line between the end points, it will have a constant acceleration

$$a = g\cos\alpha = g\frac{|y_B]}{s} = \frac{g}{\sqrt{\pi^2/4+1}}\,, \tag{3.69}$$

where $\alpha$ is the angle between the line followed by the particle and the vertical line. With $T'$ as the time spent on this path, we have $s = (1/2)aT'^2$, which gives

$$T' = \sqrt{\frac{2s}{a}} = k\sqrt{\frac{2}{g}(\frac{\pi^2}{4}+1)} = \sqrt{1+\frac{4}{\pi^2}}\,T\,. \tag{3.70}$$

Numerically this gives $T' = 1.185T$, which demonstrates, in this particular case, that the motion along the cycloid indeed is faster than along the straight line.

# PART 2

# Relativity

Chapter 4

# The four-dimensional space-time

### *Problem 4.1*

Two inertial reference frames, $S$ and $S'$, are related by the Lorentz transformation

$$x' = \gamma(x - vt)\,, \quad t' = \gamma(t - \frac{v}{c^2}x)\,, \quad y' = y\,, \quad z' = z\,. \tag{4.1}$$

a) We want to invert the transformation. In order to do so we combine the first two equations in the following ways,

$$\begin{aligned} x' + vt' &= \gamma(1 - \frac{v^2}{x^2})x = \frac{1}{\gamma}x\,, \\ t' + \frac{v}{c^2}x' &= \gamma(1 - \frac{v^2}{x^2})t = \frac{1}{\gamma}t\,. \end{aligned} \tag{4.2}$$

From this follows

$$x = \gamma(x' + vt')\,, \quad t = \gamma(t' + \frac{v}{c^2}x')\,, \tag{4.3}$$

which confirms that the inverted transformation has the same form as the original transformation, only with a sign change of the velocity, $v \to -v$.

b) We use the Lorentz transformations to relate the velocity components in $S$ and $S'$ of a moving object,

$$\begin{aligned} u'_x &= \frac{dx'}{dt'} = \frac{\gamma(dx - vdt)}{\gamma(dt - \frac{v}{c^2}dx)} = \frac{u_x - v}{1 - \frac{vu_x}{c^2}}\,, \\ u'_y &= \frac{dy'}{dt'} = \frac{dy}{\gamma(dt - \frac{v}{c^2}dx)} = \frac{1}{\gamma}\frac{u_y}{1 - \frac{vu_x}{c^2}}\,, \\ u'_z &= \frac{dz'}{dt'} = \frac{dz}{\gamma(dt - \frac{v}{c^2}dx)} = \frac{1}{\gamma}\frac{u_z}{1 - \frac{vu_x}{c^2}}\,. \end{aligned} \tag{4.4}$$

c) The following values are now assumed: The relative velocity of $S'$ and $S$ is $v = 0.5c$, the velocity of the object, as measured in $S'$, is $u' = 0.8c$, and the angle of the velocity vector $\mathbf{u}'$ is $\theta' = 45°$ relative to the $x'$ and $y'$ axes. This implies that the vector $\mathbf{u}'$ lies in the $x', y'$-plane. Thus,

$u'_z = u_z = 0$, and for the two other components we have, $u'_x = u'_y = (0.8/\sqrt{2})c$. The gamma factor for the transformation between $S$ and $S'$ is $\gamma = (1 - (v/c)^2)^{-1/2} = 1.155$.

The corresponding velocity components in $S$ are

$$\begin{aligned} u_x &= \frac{u'_x + v}{1 + \frac{vu'_x}{c^2}} = \frac{(0.8/\sqrt{2}) + 0.5}{1 + 0.5(0.8/\sqrt{2})} c = 0.83c\,, \\ u_y &= \frac{1}{\gamma} \frac{u'_y}{1 + \frac{vu'_x}{c^2}} = 0.866 \frac{0.8/\sqrt{2}}{1 + 0.5(0.8/\sqrt{2})} c = 0.38c, \end{aligned} \tag{4.5}$$

where the expressions in (4.4) have been used, with $v \to -v$ due to the inverted direction of the transformation. This gives

$$u = \sqrt{u_x^2 + u_y^2} = 0.91c\,, \tag{4.6}$$

and the angle of the velocity, relative to the $x$-axis,

$$\theta = \arctan\left(\frac{u_y}{u_x}\right)\frac{180^\circ}{\pi} = \arctan\left(\frac{u'_y}{\gamma(u'_x + v)}\right)\frac{180^\circ}{\pi} = 24.7^\circ\,. \tag{4.7}$$

In a non-relativistic treatment we see that the denominators $1 + \frac{vu'_x}{c^2}$ in the above expressions would be replaced by 1, and $\gamma$ replaced by 1. This would give a superluminal velocity of the object as seen in $S$,

$$u = \sqrt{(0.8/\sqrt{2}) + 0.5)^2 + (0.866 \cdot 0.8/\sqrt{2}))^2} = 1.21c\,. \tag{4.8}$$

The ratio between the velocity components would be modified by the change of the gamma factor,

$$\theta = \arctan\left(\frac{u'_y}{u'_x + v}\right)\frac{180^\circ}{\pi} = 28.0^\circ\,. \tag{4.9}$$

***Problem 4.2***

We examine here a combination of two boosts, which both mix the $x$ and $t$ coordinates. Expressed as $2 \times 2$ matrices they have the form,

$$L = \begin{pmatrix} \gamma & -\beta\gamma \\ -\beta\gamma & \gamma \end{pmatrix} = \begin{pmatrix} \cosh\chi & -\sinh\chi \\ -\sinh\chi & \cosh\chi \end{pmatrix}\,, \tag{4.10}$$

where the parameter $\chi$ is the rapidity. The problem to be solved is to show that in a combination of two such boosts the rapidity acts additively.

We show this by an explicit evaluation of the product, where we make use of the properties of the hyperbolic functions,

$$\begin{aligned}
L &= L_2 L_1 \\
&= \begin{pmatrix} \cosh\chi_2 & -\sinh\chi_2 \\ -\sinh\chi_2 & \cosh\chi_2 \end{pmatrix} \begin{pmatrix} \cosh\chi_1 & -\sinh\chi_1 \\ -\sinh\chi_1 & \cosh\chi_1 \end{pmatrix} \\
&= \begin{pmatrix} \cosh\chi_1\cosh\chi_2 + \sinh\chi_1\sinh\chi_2 & -\sinh\chi_1\cosh\chi_2 - \cosh\chi_1\sinh\chi_2 \\ -\sinh\chi_1\cosh\chi_2 - \cosh\chi_1\sinh\chi_2 & \cosh\chi_1\cosh\chi_2 + \sinh\chi_1\sinh\chi_2 \end{pmatrix} \\
&= \begin{pmatrix} \cosh(\chi_1+\chi_2) & -\sinh(\chi_1+\chi_2) \\ -\sinh(\chi_1+\chi_2) & \cosh(\chi_1+\chi_2) \end{pmatrix} \\
&\equiv \begin{pmatrix} \cosh\chi & -\sinh\chi \\ -\sinh\chi & \cosh\chi \end{pmatrix}, \qquad (4.11)
\end{aligned}$$

where $\chi = \chi_1 + \chi_2$. We have here used the composition rules for the hyperbolic functions

$$\begin{aligned}
\cosh(\chi_1+\chi_2) &= \cosh\chi_1\cosh\chi_2 + \sinh\chi_1\sinh\chi_2\,, \\
\sinh(\chi_1+\chi_2) &= \sinh\chi_1\cosh\chi_2 + \cosh\chi_1\sinh\chi_2\,,
\end{aligned} \qquad (4.12)$$

which are straight forward to check using the definition of these functions.

As shown, the composition rule for the Lorentz transformations gives the simple additive rule for the rapidity, $\chi = \chi_1 + \chi_2$. This is to be compared with the composition rule for the corresponding velocities.

$$v = \frac{v_1 + v_2}{1 + \frac{v_1 v_2}{c^2}}\,. \qquad (4.13)$$

### *Problem 4.3*

A thin rigid rod has rest length $L_0$. It moves relative to an inertial reference frame $S'$, so that the midpoint $A$ of the rod has the time dependent coordinates $x'_A = 0, y'_A = ut', z'_A = 0$, with $u$ as the velocity of the rod. In this reference frame the rod is at all times parallel to the $x'$-axis.

a) Since the motion of the rod, as seen in $S'$, is in the $y'$-direction, the length of the rod in the $x'$-direction is the same as in the rest frame of the rod. This implies that the end point $B$ has the $x'$ coordinate $x'_B = L_0/2$ (or alternatively $-L_0/2$). Since the rod is at all times parallel to the $x'$-axis, we have $y'_B = y'_A = ut'$ and $z'_B = z'_A = 0$.

b) We consider the Lorentz transformations between the reference frames $S'$ and $S$, with $S'$ moving with velocity $v$ along the $x$-axis relative to $S$. We are interested in finding the space coordinates in $S$ expressed as functions of the time coordinate $t$ in the same reference frame. Since the time coordinate $t'$ will then depend on the space coordinates, we consider the two points $A$ and $B$ on the rod separately.

Lorentz transformation of the coordinates at $A$ gives

$$\begin{aligned} t &= \gamma(t' + \frac{v}{c^2}x'_A) = \gamma t' , \\ x_A &= \gamma(x'_A + vt') = \gamma v t' , \\ y_A &= y'_A = ut' , \\ z_A &= z'_A = 0 . \end{aligned} \tag{4.14}$$

The first equation shows that the time coordinate in this case is $t' = t/\gamma$. Inserted in the other equations we find

$$x_A = vt , \quad y_A = \frac{1}{\gamma}ut , \quad z_A = 0 . \tag{4.15}$$

Lorentz transformations of the coordinates at $B$ give

$$\begin{aligned} t &= \gamma(t' + \frac{v}{c^2}x'_B) = \gamma(t' + \frac{v}{c^2}\frac{L_0}{2}) , \\ x_B &= \gamma(x'_B + vt') = \gamma(\frac{L_0}{2} + vt') , \\ y_B &= y'_B = ut' , \\ z_B &= z'_B = 0 . \end{aligned} \tag{4.16}$$

In this case we have $t' = \frac{1}{\gamma}t - \frac{v}{c^2}\frac{L_0}{2}$, which inserted in the equations for the space coordinates gives

$$\begin{aligned} x_B &= vt + \gamma(1 - \frac{v^2}{c^2})\frac{L_0}{2} = vt + \frac{1}{\gamma}\frac{L_0}{2} , \\ y_B &= \frac{1}{\gamma}ut - \frac{uv}{c^2}\frac{L_0}{2} , \\ z_B &= 0 . \end{aligned} \tag{4.17}$$

c) The angle $\phi$ of the rod relative to the $x$-axis in $S$ is

$$\tan\phi = \frac{y_B - y_A}{x_B - x_A} = -\frac{uvL_0/(2c^2)}{L_0/(2\gamma)} = -\frac{1}{\gamma}\frac{uv}{c^2} . \tag{4.18}$$

This shows that the rod, as seen in reference frame $S$, is tilted relative to the $x$-axis.

d) The velocity of the rod in reference $S$ has the two vector components

$$v_x = \dot{x}_A = \dot{x}_B = v\,,$$
$$v_y = \dot{y}_A = \dot{y}_B = \frac{1}{\gamma}u\,. \tag{4.19}$$

### *Problem 4.4*

We consider here a situation where a railway carriage is moving in a straight line with constant velocity $v$ relative to the earth. The earth is considered as an inertial reference frame $S$, and in this reference frame the moving carriage has the length $L$. $A$ and $B$ denote points on the rear wall and front wall of the carriage, respectively. C is a point in the middle of the carriage.

a) The situation is illustrated in the Minkowski diagram shown in Fig. 4.1 below. In the diagram the points $A$, $B$, and $C$ form parallel lines, as they represent the worldlines of points which are fixed in the carriage, and therefore move with the same speed relative to the earth. The angle between each of these lines and the time axis is determined as

$$\tan\alpha = \frac{1}{c}\frac{dx}{dt} = \frac{v}{c}\,. \tag{4.20}$$

At a given time $t_0 = 0$ a flash tube is discharged at point $C$. This event is referred to as $E_0$. The points $E_1$ and $E_2$ represent events where the light signals hit the rear wall and front wall of the carriage, respectively. The light is reflected from $A$ and $B$, and the two reflected light signals meet at a space-time point $E_3$.

b) In the Minkowski diagram the light signals and the events are shown in the following way. The event $E_0$ is the point where the worldline $C$ crosses the $x$-axis. As light moves with the speed $c$, the light signals are described in the diagram with (dotted) lines, which are tilted with angles $\theta = \pm 45°$ relative to the $x$-axis. The events $E_1$ and $E_2$ lie where the light signals hit the ends of the carriage, which means where the light paths reach the worldlines $A$ and $B$. The light signals are reflected back from these points to meet at the event $E_3$.

c) In the co-moving reference frame $S'$ of the carriage, the two events $E_1$ and $E_2$ will be simultaneous, since the distance traveled by the light signals from point $C$, in the middle of the carriage, to the rear and front ends of the carriage is the same. For the same reason the reflected light

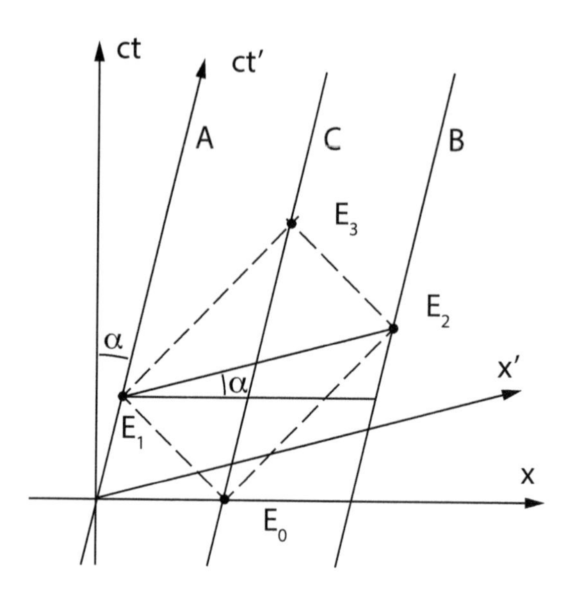


Fig. 4.1 Space-time diagram for the moving train carriage with the light signals.

signals will meet in the middle of the carriage. This is consistent with the drawing, with the event $E_3$ placed on the worldline $C$.

d) The coordinate axes of the co-moving reference frame $S'$ are included in the diagram. Since $E_1$ and $E_2$ are simultaneous in the reference frame $S'$, the line between the two is parallel to the $x'$-axis of this reference frame. Similarly the lines $A$, $B$ and $C$, which describe points which are fixed relative to the carriage, are parallel to the $ct'$-axis of $S'$. The time axis in $S'$ is tilted relative to the time axis in $S$ with the same angle (up to a sign) as the space $x'$-axis is tilted relative to the $x$-axis. This implies that the angle between the line $E_1 - E_2$ and the $x$-axis is the same as the angle $\alpha$ between the line $A$ and the $ct$-axis, as shown in the diagram.

e) Lorentz transformations between the two reference frames gives for the relative coordinates of two spacetime points $E_1$ and $E_2$,

$$\Delta t' = \gamma(\Delta t - \frac{v}{c^2}\Delta x)\,. \tag{4.21}$$

If the two points are simultaneous in $S'$, we have $\Delta t' = 0$, which implies

$$\frac{\Delta x}{\Delta t} = \frac{c^2}{v} > c\,. \tag{4.22}$$

This will be the (superluminal) velocity for an object which follows the line in Minkowski space between $E_1$ and $E_2$.

# Chapter 5

# Consequences of the Lorentz transformations

***Problem 5.1***

In this problem we study relativistic effects in a rotating disk. The radius of the disk is $R$ and the angular velocity is $\omega$.

a) A small piece of the disk has length $dr$ in the radial direction and $rd\theta$ in the angular direction, both lengths measured in the lab frame, where the center of the disk is at rest. The instantaneous inertial rest frame of the small piece moves in the angular direction with the velocity of the small piece. In the lab frame, the tangential length of this piece will be Lorentz contracted compared to length measured in the co-moving frame. In the radial direction there is no similar contraction effect. The lengths measured on the rotating disk therefore are

$$ds_r = dr\,, \quad ds_\theta = \gamma_r r d\theta\,, \tag{5.1}$$

where $\gamma_r$ is the $r$-dependent, relativistic gamma factor

$$\gamma_r = \frac{1}{\sqrt{1-\frac{r^2\omega^2}{c^2}}}\,. \tag{5.2}$$

b) The integrated length around the circumference of the disk, when measured on the rotating disk, is then

$$s_\theta = \frac{2\pi R}{\sqrt{1-\frac{R^2\omega^2}{c^2}}}\,, \tag{5.3}$$

while the radial distance measured on the disk is unchanged, $s_R = R$. The ratio between the length of the circumference and the radius, as measured on the rotating disk is therefore

$$\frac{s_\theta}{s_R} = \frac{2\pi}{\sqrt{1-\frac{R^2\omega^2}{c^2}}}\,. \tag{5.4}$$

c) Compared to the clock of the lab frame the clocks attached to the rotating disk appear to go slower by the local time dilatation factor $1/\gamma_r$.

This implies that at time $t$ of the lab frame the clock at radial coordinate $r$ will show the time

$$t_r = \sqrt{1 - \frac{r^2\omega^2}{c^2}}\, t\,. \tag{5.5}$$

d) We consider an infinitesimal spacetime displacement along the edge of the disk. With $dt$ as the time interval measured in the lab frame, $dt'$ as the time interval in the instantaneous rest frame of the disk, and $dx'$ as the displacement along the edge, measured in the same reference frame, the Lorentz transformation between the reference frames gives

$$dt = \gamma_R(dt' + \frac{\omega R}{c^2} dx')\,. \tag{5.6}$$

Assuming the spacetime displacement relates simultaneous events in the local, co-moving frame, we have $dt' = 0$, and as follows from a), $dx' = \gamma_R R d\theta$. This gives

$$dt = \gamma_R^2 \frac{\omega R^2}{c^2} d\theta\,, \tag{5.7}$$

and integrated around the full circle the result is

$$\Delta t = \gamma_R^2 \frac{2\pi\omega R^2}{c^2}\,. \tag{5.8}$$

This is the time jump measured in the lab frame, for a spacetime curve around the edge of the disk, which connects events that are locally simultaneous in the co-moving reference frame. By use of the time dilatation formula we then obtain for the time jump measured by the local, co-moving clocks,

$$\Delta t' = \frac{1}{\gamma_R}\Delta t = \gamma_R \frac{2\pi\omega R^2}{c^2} = \frac{2\pi\omega R^2}{c^2\sqrt{1 - \frac{R^2\omega^2}{c^2}}}\,. \tag{5.9}$$

### *Problem 5.2*

A spacecraft moves with the velocity $v = 3c/5$ to a point located at the distance $D = 30$ light days from the earth. After a short stop it returns with the same speed to earth. The short periods of acceleration are neglected in the following.

a) The total time measured on earth is

$$T_e = \frac{2D}{v} = \frac{10}{3}\frac{D}{c} = 100 \text{ days}\,. \tag{5.10}$$

The total time measured on the spacecraft is equal to the proper time of the spacecraft during the travel. We use the general formula for the proper time $\Delta\tau = \Delta t/\gamma$, with the gamma factor given by

$$\gamma = \frac{1}{\sqrt{1 - v^2/c^2}} = \frac{1}{\sqrt{1 - 9/25}} = \frac{5}{4}\,. \tag{5.11}$$

This, gives as the proper time of the travel,

$$T_s = \frac{1}{\gamma}T_e = \frac{4}{5}T_e = 80 \text{ days}\,. \tag{5.12}$$

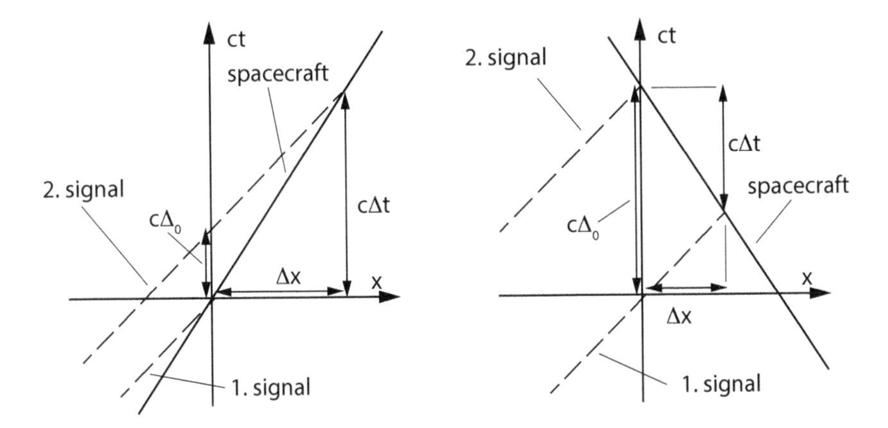


Fig. 5.1 Space-time diagrams for the radio signals received by the spacecraft, to the left, for the travel away from earth, and to the right for the travel back.

b) Every hour a signal is sent from earth to the space craft. We consider two events, where subsequent signals are received on the space craft during the travel out. Measured on earth, the coordinate difference between these events is given by (see diagram to the left in Fig. 5.1),

$$\Delta x = v\Delta t = c(\Delta t - \Delta_0) \quad \Rightarrow \quad \Delta t = \frac{c}{c - v}\Delta_0 = \frac{5}{2}\Delta_0\,. \tag{5.13}$$

The time difference between the events, measured on the spacecraft is then, due to time dilatation,

$$\Delta_1 = \frac{1}{\gamma}\Delta t = \frac{4}{5}\frac{5}{2}\Delta_0 = 2\Delta_0 = 2 \text{ hours}\,. \tag{5.14}$$

On the travel back the corresponding expressions are (see diagram to the right in Fig. 5.1),

$$\Delta x = v\Delta t = c(\Delta_0 - \Delta t) \quad \Rightarrow \quad \Delta t = \frac{c}{c + v}\Delta_0 = \frac{5}{8}\Delta_0, \tag{5.15}$$

with the time difference measured on the spacecraft as

$$\Delta_2 = \frac{1}{\gamma}\Delta t = \frac{4}{5}\frac{5}{8}\Delta_0 = \frac{1}{2}\Delta_0 = 30\,\text{min}\,. \tag{5.16}$$

c) A Minkowski diagram which shows the signals from earth to the spacecraft is shown as the diagram to the left in Fig. 5.2.

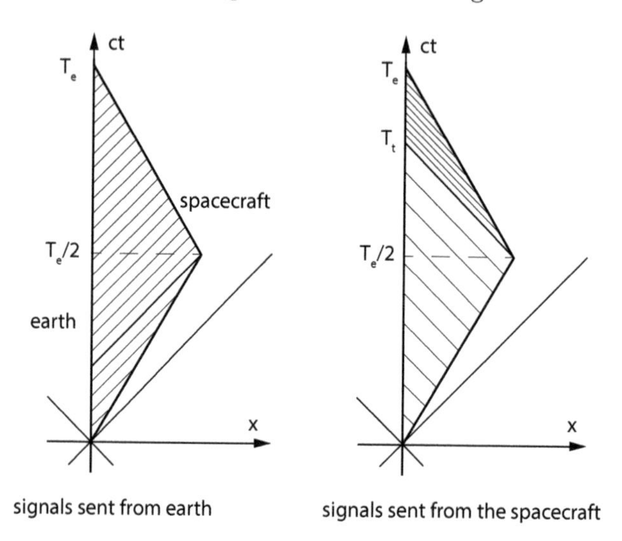


Fig. 5.2 Minkowski diagrams with the worldlines of radio signals sent from earth to the spaceship (to the left) and radio signals sent from the spacecraft to earth (to the right).

d) The situations with respect to the time intervals of the signals at the receivers are fully symmetric, whether the signals are sent from the space craft or from the earth. Only the relative velocity between the emitter and the receiver matters. This explains why the signals received on earth, on the first part of the spacecrafts travel out is $\Delta_1$ and on the second part is $\Delta_2$.

The time when the intervals of the signals received on earth change is the time when the signal sent from the space craft at halftime reaches the earth,

$$T_t = \frac{1}{2}T_e + \frac{D}{c} = \frac{D}{v} + \frac{D}{c} = \frac{D}{c}(\frac{1}{\beta} + 1) = \frac{8}{3}\frac{D}{c} = 80\,\text{days}\,. \tag{5.17}$$

e) The Minkowski diagram with the signals sent from the spacecraft to earth is shown to the right in the figure.

### *Problem 5.3*

A particle is circulating with constant speed in an accelerator ring of radius $R = 10\,\text{m}$. The speed of the particle corresponds to a relativistic gamma factor $\gamma = 100$. The laboratory frame $S$ is the rest frame of the accelerator ring, and we assume the ring to lie in the $x, y$-plane, with the center of the ring at the origin.

a) To determine the velocity $v$ of the circulating particle, with relativistic gamma factor $\gamma = 100$, we use the relations

$$\beta = v/c\,, \quad \gamma = \frac{1}{\sqrt{1-\beta^2}} \quad \Rightarrow \quad \beta = \sqrt{1 - \frac{1}{\gamma^2}}\,. \tag{5.18}$$

This gives

$$v/c = \sqrt{1 - \frac{1}{\gamma^2}} \approx (1 - \frac{1}{2\gamma^2}) = 0.99995\,. \tag{5.19}$$

The velocity $v$ is very close to the speed of light and can be put equal to c in the expressions to follow. The period of circulation in the accelerator ring is

$$T = \frac{2\pi R}{v} \approx \frac{2\pi R}{c} = 2\pi \frac{10}{3.0 \cdot 10^8} s = 2.1 \cdot 10^{-7}\,\text{s}\,, \tag{5.20}$$

and the angular velocity is

$$\omega = \frac{2\pi}{T} = \frac{c}{R} = 3.0 \cdot 10^7\,\text{s}^{-1}\,. \tag{5.21}$$

The proper time $\tau$ is the time measured on an imagined co-moving clock. It is related to the time $t$ on a coordinate clock by the time dilatation factor $1/\gamma$,

$$\tau = t/\gamma \quad \Rightarrow \quad T_\tau = T/\gamma = 2.1 \cdot 10^{-9}\,\text{s}\,, \tag{5.22}$$

with $T_\tau$ as the period of circulation measured in proper time.

b) An instantaneous inertial rest frame of the particle is an inertial reference frame where the particle has zero velocity at a specific instant. We denote by $S'$ the instantaneous inertial rest frame of the particle when it passes the point $(x, y) = (0, -R)$ of the accelerator ring. This event is assumed to correspond to vanishing time coordinates in both the lab frame $S$ and the rest frame $S'$, $t = t' = 0$. The Lorentz transformations between the two reference frames then take the form,

$$x' = \gamma(x - vt)\,, \quad y' = y + R\,, \quad t' = \gamma(t - \frac{v}{c^2}x)\,, \tag{5.23}$$

with the origin of the coordinate system of $S'$ here placed at the position of the particle.

c) In $S$, the coordinates of the accelerator ring satisfy the circular condition

$$x^2 + y^2 = R^2 \,. \tag{5.24}$$

The condition $t' = 0$ gives the following relation when applied to the coordinates of $S$,

$$t' = 0 \quad \Rightarrow \quad t = \frac{v}{c^2} x \,. \tag{5.25}$$

When inserted in the transformation formula (5.23) this gives

$$x' = \frac{1}{\gamma} x \quad \Rightarrow \quad x = \gamma x' \,. \tag{5.26}$$

With this and $y = y' - R$ introduced in the circular condition (5.24), this gives the following form for the accelerator ring in $S'$, at time $t' = 0$,

$$\gamma^2 x'^2 + (y' - R)^2 = R^2 \quad \Rightarrow \quad \frac{x'^2}{(R/\gamma)^2} + \frac{(y' - R)^2}{R^2} = 1 \,. \tag{5.27}$$

The last equation can be identified as ellipse equation where the length of the long half axis, in the $y'$-direction is $R$, and the length of the short half axis, in the $x'$-direction, is $R/\gamma$. This result is consistent with the accelerator ring being seen in $S'$ as Lorentz contracted in the direction of the relative motion between the ring and the reference frame.

d) The particle trajectory as described in the coordinates of $S$ is

$$\begin{aligned} x &= R \sin \omega t = R \sin(\gamma \omega \tau) \,, \\ y &= -R \cos \omega t = -R \cos(\gamma \omega \tau) \,, \\ t &= \gamma \tau \,. \end{aligned} \tag{5.28}$$

When re-expressed in the coordinates of $S'$, we find

$$\begin{aligned} x' &= \gamma(x - vt) = \gamma(R \sin(\gamma \omega \tau) - \gamma \omega R \tau) \,, \\ y' &= y + R = R(1 - \cos(\gamma \omega t)) \,, \\ t' &= \gamma(t - \frac{v}{c^2} x) = \gamma^2 \tau - \gamma \frac{\omega R^2}{c^2} \sin(\gamma \omega t) \,. \end{aligned} \tag{5.29}$$

A graphical representation of the trajectory in the $x', y'$-plane is shown in Fig. 5.3.

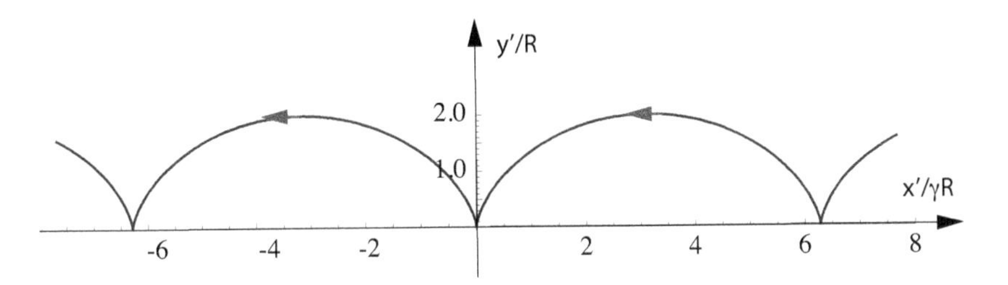


Fig. 5.3 The particle trajectory as viewed in the moving reference frame $S'$.

### *Problem 5.4*

In a particular inertial reference frame $S$ the coordinates of a spaceship are given as

$$t = \frac{c}{a_0}\sinh(\frac{a_0}{c}\tau), \quad x = \frac{c^2}{a_0}\cosh(\frac{a_0}{c}\tau), \quad y = z = 0\,, \tag{5.30}$$

with $a_0$ as a constant and $\tau$ as a time parameter.

a) The proper time $\tau$ of the space ship is related to the coordinate time $t$ by the formula

$$\Rightarrow \quad d\tau = \frac{1}{\gamma}dt = \sqrt{1 - \frac{v^2}{c^2}}dt\,, \tag{5.31}$$

where $v = dx/dt$ is the velocity of the space ship in the arbitrarily chosen reference frame. The formula is valid for any infinitesimal section of the spaceship's worldline.

To check that the parameter $\tau$ in (5.30) satisfies this condition, we differentiate the expressions given there for $t$ and $x$,

$$dt = \cosh(\frac{a_0}{c}\tau)d\tau\,, \quad dx = c\sinh(\frac{a_0}{c}\tau)d\tau. \tag{5.32}$$

This gives

$$dx^2 - c^2dt^2 = c^2(\sinh^2(\frac{a_0}{c}\tau) - \cosh^2(\frac{a_0}{c}\tau))d\tau^2 = -c^2d\tau^2\,, \tag{5.33}$$

which is equivalent to the condition (5.31).

b) A space station has $x$-coordinate $d = c^2/a_0$ and is at rest in reference frame $S$. It sends radio messages to the spaceship at regular intervals $t_n, n = 0, 1, 2, ....$ The message sent from the space station at time $t_n$ is received at the spaceship at a later time denoted $\tilde{t}_n$. Both refer to time measured in $S$. Since the message propagates with the speed of light, we have the following relation between the $x$ and $t$ coordinates of the spaceship when the message is received,

$$x(\tilde{t}_n) - d = c(\tilde{t}_n - t_n)\,. \tag{5.34}$$

Expressing this in terms of the proper time $\tau_n$ when the message is received, we get

$$\frac{c^2}{a_0}(\cosh(\frac{a_0}{c}\tau_n) - 1) = \frac{c^2}{a_0}\sinh(\frac{a_0}{c}\tau_n) - ct_n\,, \tag{5.35}$$

and from this follows

$$\cosh(\frac{a_0}{c}\tau_n) = \sinh(\frac{a_0}{c}\tau_n) - (\frac{a_0}{c}t_n - 1)\,. \tag{5.36}$$

Since $\cosh(\frac{a_0}{c}\tau_n) \geq \sinh(\frac{a_0}{c}\tau_n)$ this implies that the equation can be satisfied only as long as $t_n < c/a_0 = t_{max}$. In the limit $t_n \to t_{max}$, we clearly have $\tau_n \to \infty$ for the proper time when the message is received.

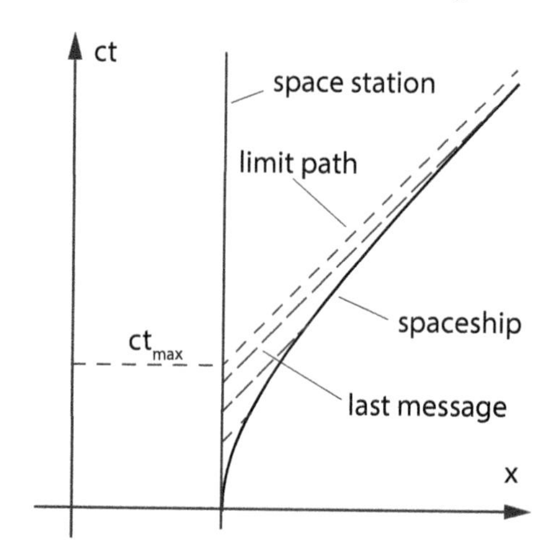


Fig. 5.4 Minkowski diagram with the hyperbolic worldline of the spacecraft, and with radio signals sent from the space station. Only signals sent before $t_{max}$ will reach the spacecraft (as long as the hyperbolic path is followed).

c) Figure 5.4 shows a Minkowski diagram with the worldlines of the spaceship and the space station in the coordinate system of reference frame $S$. The worldline of messages sent from the space station are also shown. Due to the hyperbolic form of the spaceship's worldline, only messages sent before $t_{max}$ will reach the spaceship.

Chapter 6

# Four-vector formalism and covariant equations

### *Problem 6.1*

Two four-vectors $\underline{\mathbf{A}}$ and $\underline{\mathbf{B}}$ satisfy the orthogonal relation

$$\underline{\mathbf{A}} \cdot \underline{\mathbf{B}} = \mathbf{A} \cdot \mathbf{B} - A^0 B^0 = 0\,, \tag{6.1}$$

with $\underline{\mathbf{A}}$ as a timelike vector, $\underline{\mathbf{A}}^2 = \mathbf{A}^2 - (A^0)^2 < 0$.
We will show that this implies that $\underline{\mathbf{B}}$ is a spacelike vector.

Since $\underline{\mathbf{A}}$ is a timelike vector we have

$$|\mathbf{A}| < |A^0| \quad \Rightarrow \quad |\frac{\mathbf{A}}{A^0}| < 1\,. \tag{6.2}$$

The orthogonality relation $\underline{\mathbf{A}} \cdot \underline{\mathbf{B}} = 0$ implies

$$B^0 = \frac{\mathbf{A}}{A^0} \cdot \mathbf{B}\,. \tag{6.3}$$

From this follows

$$|B^0| \le |\frac{\mathbf{A}}{A^0}|\,|\mathbf{B}| < |\mathbf{B}|\,, \tag{6.4}$$

which shows that $\underline{\mathbf{B}}$ is spacelike,

$$\underline{\mathbf{B}}^2 = \mathbf{B}^2 - (B^0)^2 > 0\,. \tag{6.5}$$

### *Problem 6.2*

a) We are given the following set of equations,

$$C^\mu = T^\mu{}_\nu A^\mu\,, \quad D_\nu = T^\mu{}_\nu A_\mu\,, \quad E_{\mu\nu\rho} = T_{\mu\nu} S^\nu{}_\rho\,, \quad G = S_{\mu\nu} T^\nu_a A^\alpha\,, \tag{6.6}$$

and the problem is to identify and correct those which do not satisfy the conditions of covariance.

We find that only the equation $D_\nu = T^\mu{}_\nu A_\mu$ is correct, while the others are

$$C^\mu = T^\mu{}_\nu A^\nu\,, \quad E_{\mu\rho} = T_{\mu\nu} S^\nu{}_\rho\,, \quad G_\mu = S_{\mu\nu} T^\nu{}_\alpha A^\alpha\,. \tag{6.7}$$

b) Starting with a set of two four-vectors and a tensor, $A^\mu$, $B^\mu$, $T^{\mu\nu}$, new scalars and vectors can be formed as follows:

$$\begin{aligned} &\bullet \textit{ scalars } A_\mu A^\mu,\ A_\mu B^\mu,\ T^\mu{}_\mu,\ T^{\mu\nu}T_{\mu\nu},\ T^{\mu\nu}A_\mu B_\nu, \ldots \\ &\bullet \textit{ vectors } T^{\mu\nu}A_\mu,\ T^{\mu\nu}T_{\nu\rho}B^\rho,\ \ldots . \end{aligned} \tag{6.8}$$

c) By manipulating $L_\mu{}^\rho L^\mu{}_\sigma$, with raising and lowering operators, and making use of the identity

$$g_{\mu\nu}L^\mu{}_\rho L^\nu{}_\sigma = g_{\rho\sigma}\,, \tag{6.9}$$

we obtain

$$\begin{aligned} L_\mu{}^\rho L^\mu{}_\sigma &= g_{\mu\alpha}L^\alpha{}_\beta\, g^{\beta\rho} L^\mu{}_\sigma \\ &= g^{\beta\rho}(g_{\mu\alpha}L^\mu{}_\sigma\, L^\alpha{}_\beta) \\ &= g_{\sigma\beta}\, g^{\beta\rho} = \delta^\rho_\sigma\,. \end{aligned} \tag{6.10}$$

***Problem 6.3***

The functions to differentiate are

$$f(x) = x_\mu x^\mu\,, \quad a^\mu(x) = x^\mu\,, \quad b^{\mu\nu}(x) = x^\mu x^\nu\,, \quad h^\mu(x) = \frac{x^\mu}{x_\nu x^\nu}\,. \tag{6.11}$$

In the following we freely change the names of repeated indices, in order to avoid unintended reuse of index names in the same equation. We evaluate the following derivatives,

$$\begin{aligned} \partial_\mu f(x) &= \frac{\partial}{\partial x^\mu}(x_\alpha x^\alpha) = \frac{\partial}{\partial x^\mu}(g_{\alpha\beta}x^\alpha x^\beta) \\ &= g_{\alpha\beta}(\delta^\alpha_\mu x^\beta + x^\alpha \delta^\beta_\mu) \\ &= g_{\mu\beta}x^\beta + g_{\alpha\mu}x^\alpha \\ &= 2x_\mu\,, \end{aligned} \tag{6.12}$$

$$\partial_\mu a^\mu(x) = \frac{\partial x^\mu}{\partial x^\mu} = \delta^\mu_\mu = 4\,, \tag{6.13}$$

$$\partial_\mu b^{\mu\nu}(x) = \frac{\partial(x^\mu x^\nu)}{\partial x^\mu} = \delta^\mu_\mu x^\nu + \delta^\nu_\mu x^\mu = 4x^\nu + x^\nu = 5x^\nu\,, \tag{6.14}$$

$$
\begin{aligned}
\partial_\mu h^\mu(x) &= \frac{\partial}{\partial x^\mu}\left(\frac{x^\mu}{x_\nu x^\nu}\right) \\
&= \frac{4}{x_\nu x^\nu} - \frac{x^\mu}{(x_\nu x^\nu)^2} g_{\alpha\beta}\frac{\partial}{\partial x^\mu}(x^\alpha x^\beta) \\
&= \frac{4}{x_\nu x^\nu} - \frac{x^\mu}{(x_\nu x^\nu)^2} g_{\alpha\beta}(\delta^\alpha_\mu x^\beta + \delta^\beta_\mu x^\alpha) \\
&= \frac{4}{x_\nu x^\nu} - \frac{2x^\mu x_\mu}{(x_\nu x^\nu)^2} \\
&= \frac{2}{x_\nu x^\nu}\,. \qquad (6.15)
\end{aligned}
$$

### *Problem 6.4*

An inertial reference frame $S$ has time and space axes defined by the basis vectors $\underline{\mathbf{e}}_\mu$, $\mu = 0, 1, 2, 3$, with the generalized orthonormalization condition

$$\underline{\mathbf{e}}_\mu \cdot \underline{\mathbf{e}}_\nu = g_{\mu\nu}\,. \qquad (6.16)$$

A second inertial frame $S'$ has coordinate axes with unit vectors that mix those of $S$ in the following way,

$$
\begin{aligned}
\underline{\mathbf{e}}'_0 &= \cosh\chi\,\underline{\mathbf{e}}_0 + \sinh\chi\,\underline{\mathbf{e}}_1\,, \\
\underline{\mathbf{e}}'_1 &= \sinh\chi\,\underline{\mathbf{e}}_0 + \cosh\chi\,\underline{\mathbf{e}}_1\,, \qquad (6.17)
\end{aligned}
$$

while $\underline{\mathbf{e}}_2$ and $\underline{\mathbf{e}}_3$ are left unchanged.

a) We will find the relation between the parameter $\chi$ and the relative velocity $v$ between the two reference frames. In order to do so, we make use of the following relations between the spacetime coordinates and the unit vectors in the reference frame $S$,

$$\underline{\mathbf{x}} = x^\mu \underline{\mathbf{e}}_\mu \quad \Rightarrow \quad x_\mu = \underline{\mathbf{e}}_\mu \cdot \underline{\mathbf{x}}\,, \qquad (6.18)$$

with similar expressions in reference frame $S'$. The transformation equations between the two sets of unit vectors then give,

$$
\begin{aligned}
x'_0 &= \cosh\chi\, x_0 + \sinh\chi\, x_1\,, \\
x'_1 &= \sinh\chi\, x_0 + \cosh\chi\, x_1\,, \\
x'_2 &= x_2, \quad x'_3 = x_3\,. \qquad (6.19)
\end{aligned}
$$

With $x_0 = -ct$, $x_1 = x$, $x_2 = y$, $x_3 = z$, etc., this gives the coordinate transformations

$$
\begin{aligned}
ct' &= \cosh\chi\, ct - \sinh\chi\, x\,, \\
x' &= -\sinh\chi\, ct + \cosh\chi\, x\,, \\
y' &= y, \quad z' = z\,. \qquad (6.20)
\end{aligned}
$$

The equations have the standard form of a Lorentz transformation formula, where reference frame $S'$ moves with velocity $v$ along the $x$-axis relative to to reference system $S$, provided we make the identifications

$$\cosh\chi = \gamma\,, \quad \sinh\chi = \beta\gamma\,. \tag{6.21}$$

b) We assume that in a (two-dimensional) Minkowski diagram the basis vectors of reference system $S$, $(\underline{\mathbf{e}}_0, \underline{\mathbf{e}}_1)$, are treated as orthogonal and normalized to 1. The corresponding basis vectors of $S'$, which are related to the basis vectors of $S$ by $\underline{\mathbf{e}}'_0 = \cosh\chi\,\underline{\mathbf{e}}_0 + \sinh\chi\,\underline{\mathbf{e}}_1$ and $\underline{\mathbf{e}}'_1 = \sinh\chi\,\underline{\mathbf{e}}_0 + \cosh\chi\,\underline{\mathbf{e}}_1$, will not be orthogonal in the same diagram, and neither are normalized to 1. In particular the basis vector $\underline{\mathbf{e}}'_0$ will be rotated by an angle $\phi$ relative to $\underline{\mathbf{e}}_0$, with $\cosh\chi = a\cos\phi$ and $\sinh\chi = a\sin\phi$, where $a$ is a normalization factor. This gives

$$\tan\phi = \tanh\chi = \beta = v/c\,. \tag{6.22}$$

Similarly $\underline{\mathbf{e}}'_1$ will be rotated by the same angle relative to $\underline{\mathbf{e}}_1$, but in the opposite direction. For the angle $\phi = 30°$ we have

$$v = \tan 30°c = c/\sqrt{3} = 0.577c\,. \tag{6.23}$$

For $\phi = 15°$ we similarly find

$$v = \tan 15°c = \frac{\sin 30°}{\cos 30° + 1} = c/(\sqrt{3}+2) = 0.268c\,. \tag{6.24}$$

c) We express the vector $\underline{\mathbf{e}}'_0$ in terms of the basis vectors of $S$ as

$$\underline{\mathbf{e}}'_0 = \cosh\chi\,\underline{\mathbf{e}}_0 + \sinh\chi\,\underline{\mathbf{e}}_1 \equiv x^0\underline{\mathbf{e}}_0 + x^1\underline{\mathbf{e}}_1\,. \tag{6.25}$$

The coefficients satisfy

$$(x^0)^2 - (x^1)^2 = \cosh^2\chi - \sinh\chi = 1\,, \tag{6.26}$$

which show that the two parameters $x^0$ and $x^1$ define a hyperbola in the diagram. For $\underline{\mathbf{e}}'_1$ we find the same equation, but with a switch of the parameters $x^0$ and $x^1$. This means that the hyperbola in this case is obtained by reflecting the first hyperbola about the line $x^0 = x^1$.

Chapter 7

# Relativistic kinematics

*Problem 7.1*

a) An electron is moving in a storage ring of radius $R = 10\text{m}$, with a speed that corresponds to the gamma factor $\gamma = 30$. The velocity of the particle is determined by the gamma factor in the following way

$$\gamma = \frac{1}{\sqrt{1 - \frac{v^2}{c^2}}} \quad \Rightarrow \quad v = \sqrt{1 - \frac{1}{\gamma^2}}\, c\,. \tag{7.1}$$

With $\gamma = 30$ this gives

$$v = 0.9994c = 2.998 \cdot 10^8 \,\text{m/s}\,. \tag{7.2}$$

The period of circulation, measured in the lab frame, is

$$T = \frac{2\pi R}{v} = 2.0956 \cdot 10^{-7}\,\text{s} \tag{7.3}$$

and the proper time of one period is reduced by the time dilatation factor,

$$T_p = T/\gamma = 6.985 \cdot 10^{-9}\,\text{s}\,. \tag{7.4}$$

The acceleration measured in the lab frame is

$$a = \frac{v^2}{R} = 8.990 \cdot 10^{15}\,\text{m/s}^2 \tag{7.5}$$

and the corresponding proper acceleration, evaluated for circular motion with constant velocity, is

$$a_0 = \gamma^2 a = 8.091 \cdot 10^{18}\,\text{m/s}^2\,. \tag{7.6}$$

b) We now study a different system. The motion of a particle in two dimensions is described by the time dependent position vector

$$\mathbf{r} = ut\mathbf{i} + \frac{1}{2}gt^2\mathbf{j}\,, \tag{7.7}$$

with the velocity,

$$\mathbf{v} = u\mathbf{i} + gt\mathbf{j}\,, \tag{7.8}$$

and acceleration,

$$\mathbf{a} = \frac{dv}{dt} = g\mathbf{j}\,. \tag{7.9}$$

The corresponding gamma factor is

$$\gamma = \frac{1}{\sqrt{1 - \frac{u^2+g^2t^2}{c^2}}}\,. \tag{7.10}$$

We make use of the general formula for the absolute value of the proper acceleration

$$\begin{aligned}
\mathbf{a}_0^2 &= \gamma^4\mathbf{a}^2 + \gamma^6\frac{(\mathbf{v}\cdot\mathbf{a})^2}{c^2} \\
&= \frac{g^2}{(1 - \frac{u^2+g^2t^2}{c^2})^2} + \frac{g^4t^2}{c^2(1 - \frac{u^2+g^2t^2}{c^2})^3} \\
&= (\frac{1}{u^2+g^2t})^3 g^2\left((1 - \frac{u^2+g^2t^2}{c^2}) + \frac{g^2t^2}{c^2}\right) \\
&= g^2\frac{1 - \frac{u^2}{c^2}}{(1 - \frac{u^2+g^2t^2}{c^2})^3}\,.
\end{aligned} \tag{7.11}$$

This gives

$$a_0 = g\sqrt{\frac{c^2-u^2}{(c^2-u^2-g^2t^2)^3}}\,. \tag{7.12}$$

The proper acceleration is seen to increase with increasing $t$. It goes in fact to infinity when $t \to \sqrt{c^2-u^2}/g$. This corresponds to the time where the particle velocity reaches the velocity of light, $v \to c$, which is, of course, an unrealistic limit.

## *Problem 7.2*

A spacecraft leaves the earth at local time $t = 0$ and travels to the star Proxima Centauri, at a distance of $d = 4.2$ light years. The spacecraft follows a linear path, along the $x$-axis in an earth-fixed reference frame. The initial value of the position coordinate is $x = 0$, and the (proper) time coordinate measured on the spacecraft is set to $\tau = 0$ on departure.

The spacecraft follows on the first part of the journey, until it is halfway to the star, a hyperbolic path of the form

$$x - x_I = \frac{c^2}{a}\cosh(\frac{a}{c}(\tau - \tau_I)), \quad t - t_I = \frac{c}{a}\sinh(\frac{a}{c}(\tau - \tau_I))\,. \tag{7.13}$$

The same type of spacetime path is followed on the second part of the journey, until it reaches Proxima Centauri, but with the opposite sign of the parameter $a$. The journey back from the star is performed in the same way, in the opposite direction. A short stop at the star is disregarded in the description.

a) The initial conditions for part $I$ of the journey, $t = 0, x = 0, v = 0, \tau = 0$, determines the coefficients $x_I$, $t_I$ and $\tau_I$ of (7.13),

$$\begin{aligned} v &= c\tanh(\frac{a}{c}\tau_I) = 0 \quad \Rightarrow \quad \tau_I = 0, \\ x &= x_I + \frac{c^2}{a} = 0 \quad \Rightarrow \quad x_I = -\frac{c^2}{a}, \\ t &= t_I = 0. \end{aligned} \tag{7.14}$$

This gives for the spacetime coordinates of the spaceship on part $I$ of the journey,

$$x = \frac{c^2}{a}(\cosh(\frac{a}{c}\tau) - 1), \quad t = \frac{c}{a}\sinh(\frac{a}{c}\tau) . \tag{7.15}$$

b) To show that the parameter $\tau$ is the proper time of the spaceship, we consider an infinitesimal change in the space-time coordinates of the spacecraft,

$$dx = c\sinh(\frac{a}{c}\tau)d\tau\,, \quad dt = \cosh(\frac{a}{c}\tau)d\tau\,. \tag{7.16}$$

The corresponding Lorentz invariant line element is

$$ds^2 = dx^2 - c^2dt^2 = -c^2\tau^2\,, \tag{7.17}$$

which gives

$$d\tau = \sqrt{dt^2 - \frac{1}{c^2}dx^2} = \sqrt{1 - \frac{v^2}{c^2}}dt = \frac{1}{\gamma}dt\,. \tag{7.18}$$

This shows that $\tau$ satisfies the definition of proper time of the spacecraft.

The components of the four-velocity are

$$\begin{aligned} U^0 &= c\frac{dt}{d\tau} = c\cosh(\frac{a}{c}\tau) = \gamma c\,, \\ U^1 &= \frac{dx}{d\tau} = c\sinh(\frac{a}{c}\tau) = \gamma v\,, \\ U^2 &= U^3 = 0\,, \end{aligned} \tag{7.19}$$

and the components of the four-acceleration are

$$\begin{aligned} A^0 &= \frac{dU^0}{d\tau} = a\sinh(\frac{a}{c}\tau)\,, \\ A^1 &= \frac{dU^1}{d\tau} = a\cosh(\frac{a}{c}\tau)\,, \\ A^2 &= A^3 = 0\,. \end{aligned} \tag{7.20}$$

The proper acceleration is determined by

$$a_0^2 = \underline{\mathbf{A}}^2 = (A^1)^2 - (A^0)^2 = a^2 \quad \Rightarrow \quad a_0 = a\,. \tag{7.21}$$

This shows that the parameter $a$ in the definition of the coordinates of the spacecraft is identical to its proper acceleration. The proper acceleration is therefore constant, except for a sign change under parts $II$ and $III$ of the journey.

c) The Minkowski diagram with the four parts of the journey is shown in Fig. 7.1. Parts $II$ – $IV$ of the journey are generated from part $I$ by reflections in the space and time coordinates, combined with a shift in positions.

d) We assume now that the proper acceleration is equal to the acceleration of gravity on earth, $a = 9.8\,\mathrm{m/s^2}$. The proper time $\tau_0$ when the acceleration is reversed, corresponds to the position of the spacecraft being halfway to Proxima Centauri, $x = d/2$. This gives

$$\begin{aligned} &\frac{c^2}{a}(\cosh(\frac{a}{c}\tau_0) - 1) = \frac{d}{2} \quad \Rightarrow \\ &\cosh(\frac{a}{c}\tau_0) = 1 + \frac{ad}{2c^2} = 3.16\,. \end{aligned} \tag{7.22}$$

This determines $\tau_0$ as

$$\tau_0 = \frac{c}{a}\mathrm{arccosh}(3.16) = 1.76\,\mathrm{years}\,. \tag{7.23}$$

From this follows that the total proper time and the earth time spent on the full journey are

$$\begin{aligned} \tau_{tot} &= 4\tau_0 = 7.1\,\mathrm{years}\,, \\ t_{tot} &= 4\frac{c}{a}\sinh(\frac{a}{c}\tau_0) = 11.7\,\mathrm{years}\,. \end{aligned} \tag{7.24}$$

e) The spacecraft has the maximum speed at the point where the acceleration is reversed, which means when $\tau = \tau_0$,

$$v_{max} = \tanh(\frac{a}{c}\tau_0)c = 0.95c\,. \tag{7.25}$$

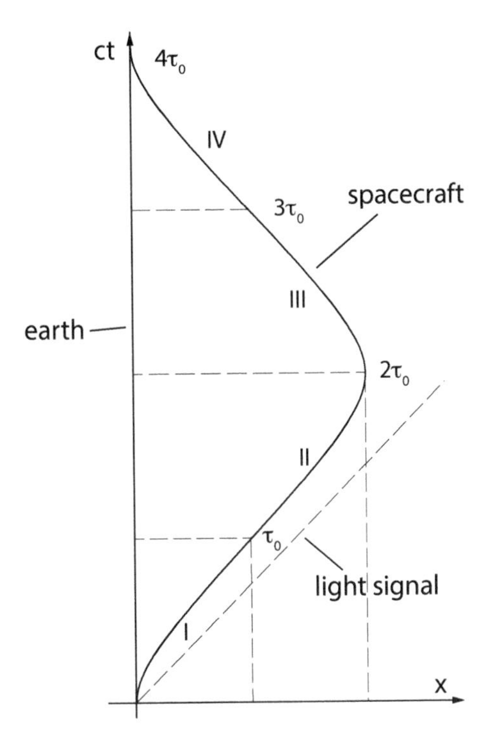


Fig. 7.1 Minkowski diagram of the space-time journey to Proxima Centauri and back.

## *Problem 7.3*

The two end points of a moving rod have the following time dependent coordinates in a reference frame $S$,

$$x_A = c\sqrt{t^2 + c^2/a^2}\,, \quad x_B = c\sqrt{t^2 + c^2/b^2}\,, \quad a > b. \tag{7.26}$$

a) A second reference frame moves in the same direction with a velocity $v$ relative to $S$. We will show that the end points of the rod satisfy the same equations when expressed in the coordinates of $S'$.

We make use of the following relations, which follow from (7.26),

$$x_A^2 - c^2t^2 = \frac{c^2}{a^2}\,, \quad x_B^2 - c^2t^2 = \frac{c^2}{b^2}\,, \tag{7.27}$$

and since $x^2 - c^2t^2$ is a Lorentz invariant, the following equality is valid for both points $A$ and $B$,

$$x^2 - c^2t^2 = x'^2 - c^2t'^2\,. \tag{7.28}$$

This implies also that the coordinates in reference frame $S'$ satisfy the relations

$$x'_A = c\sqrt{t'^2 + c^2/a^2}\,, \quad x'_B = c\sqrt{t'^2 + c^2/b^2}\,, \quad a > b\,. \tag{7.29}$$

b) The velocities of the end points $A$ and $B$, measured in $S$ are

$$\begin{aligned} v_A(t) &= \frac{dx_A}{dt} = \frac{ct}{\sqrt{t^2 + c^2/a^2}}\,, \\ v_B(t) &= \frac{dx_B}{dt} = \frac{ct}{\sqrt{t^2 + c^2/b^2}}\,. \end{aligned} \tag{7.30}$$

At time $t = 0$ this implies

$$v_A = v_B = 0\,, \tag{7.31}$$

which shows that $S$ is an instantaneous rest frame for both points $A$ and $B$ at $t = 0$. Similarly we find, for the velocities of the end points in reference frame $S'$, at time $t' = 0$

$$v'_A = v'_B = 0\,. \tag{7.32}$$

This shows that reference frame $S'$ is a common rest frame for $A$ and $B$ at $t' = 0$.

The distance between $A$ and $B$, measured in $S$ at time $t = 0$, is

$$d = x_B - x_A = \frac{c^2}{b} - \frac{c^2}{a} = c^2\frac{a-b}{a+b}\,. \tag{7.33}$$

Similarly the distance between $A$ and $B$, measured in $S'$ at time $t' = 0$ is

$$d' = x'_B - x'_A = \frac{c^2}{b} - \frac{c^2}{a} = c^2\frac{a-b}{a+b} = d\,. \tag{7.34}$$

The length measured in the two instantaneous inertial rest frames, at $t = 0$ and $t' = 0$, respectively, is therefore the same.

The velocity $v$ of the reference frame $S'$ relative to $S$ can be regarded as a free variable. It is, as shown above, identical to the velocity of the end points $A$ and $B$ at time $t' = 0$, as measured in $S$. When $v$ is continuously changed, this will correspond to moving continuously along the worldlines of the two end points. Thus, the motion of the end points is such that they will always have a common instantaneous inertial rest frame, and the length will be constant in this frame, equal to $d$. The rod will therefore move like a rigid body, in spite of the fact that the motion is relativistic.

c) The acceleration of $A$, measured in $S$, is at time $t = 0$,

$$a_A(0) = (\frac{dv_A}{dt})_{t=0} = a\,. \tag{7.35}$$

Similarly, at $t' = 0$ in $S'$,

$$a'_A(0) = (\frac{dv'_A}{dt})_{t'=0} = a\,. \tag{7.36}$$

This means that the accelerations, measured in the instantaneous inertial rest frame at different points along the worldline of $A$, all have the same value $a$. The same result applies to $B$ with $a$ replaced by $b$.

d) A light signal which is sent from $A$ at time $t = 0$ is received at $B$ at time $t_B$. We have

$$x_B(t_B) = x_A(0) + ct_B\,, \quad \Rightarrow \quad c\sqrt{t_B^2 + \frac{c^2}{b^2}} = \frac{c^2}{a} + ct_B\,. \tag{7.37}$$

The squared equation determines $t_B$,

$$c^2 t_B^2 + \frac{c^4}{b^2} = c^2 t_B^2 + 2t_B\frac{c^3}{a} + \frac{c^4}{a^2} \quad \Rightarrow \quad t_B = \frac{1}{2}c\,a(\frac{1}{b^2} - \frac{1}{a^2})\,, \tag{7.38}$$

and the coordinate $x_B$ at this instant,

$$x_B(t_B) = \frac{c^2}{a} + ct_B = \frac{1}{2}c^2 a(\frac{1}{a^2} + \frac{1}{b^2})\,. \tag{7.39}$$

The velocity of $B$ at $t_B$ is then

$$v_B(t) = \frac{ct_B}{\sqrt{t_B^2 + c^2/b^2}} = c\frac{ct_B}{x_B} = c\frac{a^2 - b^2}{a^2 + b^2}\,. \tag{7.40}$$

The light signal, which is sent with frequency $\nu_0$ from $A$ at $t = 0$, is received at $B$ with frequency $\nu_B$ determined by the Doppler formula,

$$\nu_B = \sqrt{\frac{1 - \beta_B}{1 + \beta_B}}\nu_0 = \sqrt{\frac{1 - \frac{a^2 - b^2}{a^2 + b^2}}{1 + \frac{a^2 - b^2}{a^2 + b^2}}}\nu_0 = \frac{b}{a}\nu_0\,. \tag{7.41}$$

### *Problem 7.4*

A spacecraft passes the earth with velocity $v = 0.8c$, with $d$ as the shortest distance between the spacecraft and the earth. In an earth-fixed frame $S$ the position of the spacecraft is described by coordinates

$$x(t) = vt\,, \quad y = d\,, \quad z = 0. \tag{7.42}$$

When passing, the spacecraft is continuously submitting radio messages to the earth on frequency $\nu_0$. An antenna on earth, located at the origin of the coordinate system, receives the messages and registers the frequency $\nu(t)$ and direction of the received signal during the passage. This direction is measured by the angle $\theta(t)$ between the signal and the $x$-axis.

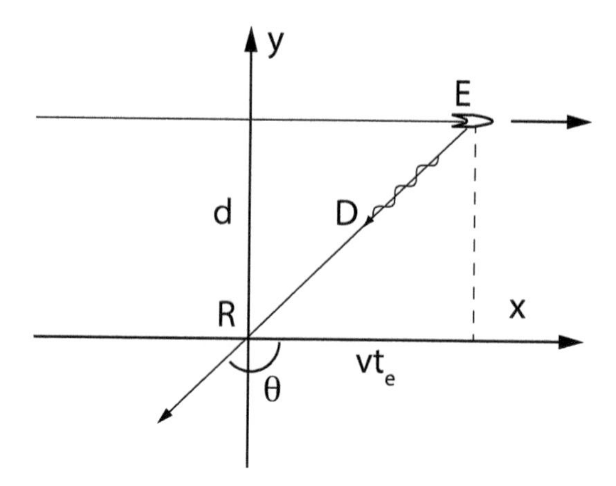


Fig. 7.2 Illustration of the emission of messages from a passing spacecraft.

a) A radio signal is received at time t on earth. It is emitted a bit earlier, at time $t_e = t - \Delta t$. We will determine $\Delta t$ as a function of $t$.

The coordinates $(ct, x, y)$ for emission $E$, and for reception $R$ of a message, are in the earth-fixed reference frame $S$ (see Fig. 7.2),

$$E: \ (ct_e, vt_e, d)\,, \quad R: \ (ct, 0, 0)\,. \tag{7.43}$$

The time difference and distance between the two events are

$$\Delta t = t - t_e\,, \quad D = \sqrt{v^2 t_e^2 + d^2}\,. \tag{7.44}$$

Since the separation of the two events is lightlike, this implies

$$c^2 \Delta t^2 = D^2 = v^2 (t - \Delta t)^2 + d^2\,, \tag{7.45}$$

which gives a second order equation for $\Delta t$,

$$(c^2 - v^2)\Delta t^2 + 2v^2 t \Delta t - v^2 t^2 - d^2 = 0\,. \tag{7.46}$$

This is rewritten as

$$\Delta t^2 + 2\beta^2\gamma^2 t \Delta t - \gamma^2(\beta^2 t^2 + d^2/c^2) = 0\,, \tag{7.47}$$

and with $\Delta t$ positive, the solution is

$$\begin{aligned} \Delta t &= -\beta^2\gamma^2 t + \sqrt{(\gamma^4\beta^4 + \gamma^2\beta^2)t^2 + \gamma^2 d^2/c^2} \\ &= -\beta^2\gamma^2 t + \sqrt{\gamma^4\beta^2 t^2 + \gamma^2 d^2/c^2}\,. \end{aligned} \tag{7.48}$$

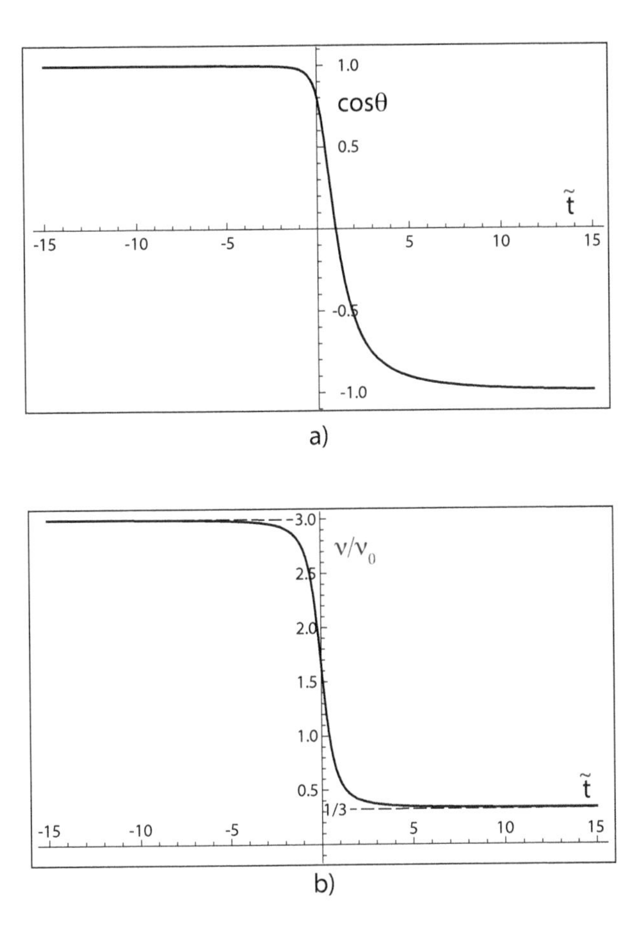


Fig. 7.3 Diagrams for the time dependence of frequency and direction of signals sent from the passing spacecraft to earth. In Panel a) is shown the change in direction of the signals, and in Panel b), the change in frequency. The dimensionless time coordinate $\tilde{t} = ct/d$ is used.

b) The angle $\theta$ of the signal when reached on earth, is determined by

$$\cos\theta = -\frac{vt_e}{D} = -\frac{v(t-\Delta t)}{\sqrt{v^2(t-\Delta t)^2 + d^2}}\,. \tag{7.49}$$

The result for $\cos\theta$ is plotted as a function of the dimensionless time coordinate $\tilde{t} = ct/d$ in the upper panel of Fig 7.3.

c) The relevant Doppler formula (see the textbook) gives, for the frequency of the signal registered in the receiver,

$$\nu = \frac{\nu_0}{\gamma(1-\beta\cos\theta)}\,. \tag{7.50}$$

The ratio $\nu/\nu_0$ is shown as a function of $\tilde{t}$ in the lower panel of the figure.

When $t \to \pm\infty$ we have $\cos\theta \to \mp 1$. The asymptotic values of the frequency $\nu$ in these limits are

$$\nu_{\pm\infty} \to \frac{\nu_0}{\gamma(1\pm\beta)} = \sqrt{\frac{1\mp\beta}{1\pm\beta}}\,\nu_0\,. \tag{7.51}$$

This gives the asymptotic values, $\nu_{-\infty} = 3\nu_0$ and $\nu_{+\infty} = \nu_0/3$.

# Chapter 8

# Relativistic dynamics

*Problem 8.1*

Two photons in the laboratory system have frequencies $\nu_1$ and $\nu_2$. The angle between the propagation directions is $\theta$.

a) The problem here is to find the total energy and the absolute value of the total momentum of the photons in the lab frame. We choose the $x$-axis in the direction of photon 1, and the $y$-axis orthogonally directed, in the plane of the two photons. The four-momenta of the two photons are then

$$\underline{\mathbf{P}}_1 = \frac{h\nu_1}{c}(1,1,0,0)\,, \quad \underline{\mathbf{P}}_2 = \frac{h\nu_2}{c}(1,\cos\theta,\sin\theta,0)\,, \tag{8.1}$$

and the total four-momentum is

$$\underline{\mathbf{P}} = \underline{\mathbf{P}}_1 + \underline{\mathbf{P}}_2 = \frac{h}{c}(\nu_1+\nu_2, \nu_1+\nu_2\cos\theta, \nu_1+\nu_2\sin\theta, 0)\,. \tag{8.2}$$

The total energy is determined by the 0th component of the four-momentum,

$$E = cP^0 = h(\nu_1+\nu_2)\,, \tag{8.3}$$

and the total momentum is determined by the three-vector part of the four-momentum. The absolute value is

$$\begin{aligned} |\mathbf{p}| &= \frac{h}{c}\sqrt{(\nu_1+\nu_2\cos\theta)^2 + \nu_2^2\sin^2\theta} \\ &= \frac{h}{c}\sqrt{\nu_1^2+\nu_2^2+2\nu_1\nu_2\cos\theta}\,. \end{aligned} \tag{8.4}$$

b) To find the photons' frequency in the CM-system we exploit the fact that the total momentum of the two photons vanishes in this reference frame, and that $\underline{\mathbf{P}}^2 = \mathbf{p}^2 - E^2/c^2$ is a Lorentz invariant. This gives the following relation between the values of the invariant in the CM-system and the lab-system,

$$\begin{aligned} E_{CM}^2 &= E^2 - c^2\mathbf{p}^2 \\ &= h^2[(\nu_1+\nu_2)^2 - \nu_1^2 - \nu_2^2 - 2\nu_1\nu_2\cos\theta)] \\ &= 2h^2\nu_1\nu_2(1-\cos\theta)\,. \end{aligned} \tag{8.5}$$

Since $E_{CM} = 2h\nu_{CM}$, with $\nu_{CM}$ as the photon frequency, which is equal for the two photons in the CM-system, we find

$$\nu_{CM} = \sqrt{\nu_1\nu_2(1-\cos\theta)} = \sqrt{2\nu_1\nu_2}\sin\left(\frac{\theta}{2}\right). \tag{8.6}$$

c) If the momenta of the two photons are collinear, both photons will propagate with the speed of light in the same direction. In this case there is no inertial reference frame where the two photons have the same momentum. Thus, the limit $\theta \to 0$ can be viewed as the limit where the velocity of the CM-system goes to the speed of light. In this limit, as the result above shows, the frequency of the photons in the CM-system goes to zero.

### *Problem 8.2*

a) The problem here is to determine the minimum energy of the photon, if the following process should be possible,

$$\gamma + e^- \to e^- + e^- + e^+, \tag{8.7}$$

with the electron being at rest before the collision with the photon.

It is convenient to consider the situation in the CM-system, where the total momentum vanishes, i.e. $\mathbf{p} = 0$. In this system the smallest energy of the three particles corresponds to the situation where all three particles are at rest. The value of the total energy is then $E_{CM} = 3m_ec^2$, with $m_e$ the electron mass. Energy conservation implies that the photon and the electron, before the collision, have the same minimum energy $E_{CM}$, in the CM-system.

We next make use of the Lorentz invariant $E^2 - p^2c^2 = E_{CM}^2$, with $E$ and $p$ as the total energy and momentum of the photon and electron in the lab system. Since the photon momentum is $h\nu/c$, the equality gives

$$(h\nu + m_ec^2)^2 - (h\nu)^2 = (3m_ec^2)^2 \quad \Rightarrow \quad h\nu = 4m_ec^2. \tag{8.8}$$

Thus, the minimum energy of the photon needed for the given process to happen is four times the rest energy of an electron.

b) One way to see that the process

$$\gamma \to e^- + e^+ \tag{8.9}$$

is impossible, is to note that since the electron and positron have masses different from zero, one can always find a CM-system for the $e^-+e^+$-system. However, since the photon is massless, a CM-system does not exist for a single photon.

The impossibility can also be seen as a conflict between conservation of energy and momentum in the given process. With the $x$-axis taken as the direction of propagation of the photon, and $\nu$ as the frequency of the photon, conservation of momentum can be written as

$$h\nu/c = p_{1x} + p_{2x}\,, \quad p_{1y} + p_{2y} = 0\,, \tag{8.10}$$

and conservation of energy as

$$h\nu = \sqrt{\mathbf{p}_1^2 c^2 + m^2c^2} + \sqrt{\mathbf{p}_2^2 c^2 + m^2c^2}\,. \tag{8.11}$$

But clearly $\sqrt{\mathbf{p}^2c^2 + m^2c^2} > |p_x|c$, and the equations for conservation of energy and momentum therefore cannot both be satisfied.

***Problem 8.3***

a) We consider the situation where two colliding particles form a single particle after the collision. The velocity of one of the particles is $v$, along the $x$-axis, while the other is at rest before the collision. The total energy and momentum of the compound particle are determined by conservation of the below quantities in the collision,

$$\begin{aligned} E &= \Gamma M c^2 = (\gamma + 1)mc^2\,, \\ P &= \Gamma M V = \gamma m v\,, \end{aligned} \tag{8.12}$$

with $\gamma = (1-v^2/c^2)^{-1/2}$ and $\Gamma = (1-V^2/c^2)^{-1/2}$, and with $V$ the velocity of the compound particle. The mass $M$ of the compound particle is determined by the relativistic energy-momentum relation, $E^2 - c^2P^2 = M^2c^2$,

$$\begin{aligned} M^2 &= (\gamma+1)^2m^2 - \gamma^2 m^2(v^2/c^2) \\ &= (\gamma^2(1 - \frac{v^2}{c^2}) + 2\gamma + 1)m^2 \\ &= 2(\gamma + 1)m^2\,, \end{aligned} \tag{8.13}$$

which gives $M = \sqrt{2(\gamma+1)}m$. The velocity $V$ of the compound particle is

$$V = \frac{c^2P}{E} = \frac{\gamma\beta}{\gamma+1}c = \sqrt{\frac{\gamma-1}{\gamma+1}}c\,. \tag{8.14}$$

b) We now assume that the two particles collide elastically. Momentum conservation gives, for the components of the particle momenta after the collision,

$$p_{1x} + p_{2x} = \gamma m v\,, \quad p_{1y} + p_{2y} = 0\,, \tag{8.15}$$

where the particle velocities are assumed to make the same angle with the $x$-axis. The conservation equation for the $y$-components of the momenta gives

$$|\mathbf{p}_1|\cos\theta = |\mathbf{p}_2|\cos\theta \quad \Rightarrow \quad |\mathbf{p}_1| = |\mathbf{p}_2| \equiv p\,. \tag{8.16}$$

From this follows that the energies of the two particles are the same

$$E_1 = E_2 = \sqrt{p^2c^2 + m^2c^4}\,. \tag{8.17}$$

c) Energy conservation gives

$$E_1 = E_2 = \frac{1}{2}(\gamma+1)mc^2\,, \tag{8.18}$$

and from this follows that the particle momentum is

$$p = \sqrt{\frac{E_1^2}{c^2} - m^2c^2} = \frac{1}{2}\sqrt{\gamma^2 + 2\gamma - 3}\,mc\,. \tag{8.19}$$

d) The angle $\theta$ is determined in the following way. The conservation of the $x$-component of the momentum gives

$$p\cos\theta = \frac{1}{2}\gamma m v \quad \Rightarrow \quad \cos\theta = \frac{\gamma\beta}{\gamma^2 + 2\gamma - 3}\,. \tag{8.20}$$

The expression can be simplified by using the identities $\beta = \sqrt{1 - (1/\gamma^2)}$ and $\gamma^2 + 2\gamma - 3 = (\gamma-1)(\gamma+3)$,

$$\cos\theta = \sqrt{\frac{\gamma^2 - 1}{\gamma^2 + 2\gamma - 3}} = \sqrt{\frac{\gamma+1}{\gamma+3}}\,. \tag{8.21}$$

In the limit $\gamma \to 1$ we have $\cos\theta = 1/\sqrt{2}$, which means $\theta = \pi/4$, and in the limit $\gamma \to \infty$ we have $\cos\theta = 1$, which means $\theta = 0$. The ratio

$$\frac{\gamma+1}{\gamma+3} = 1 - \frac{2}{\gamma+3} \tag{8.22}$$

increases monotonically with $\gamma$ in the interval $1 < \gamma < \infty$. This implies that $\theta$ decreases monotonically in the same interval, and thus is always smaller than the limit value $\theta = \pi/4$.

### *Problem 8.4*

A photon with energy $E_{ph} = 100\,\mathrm{keV}$ is scattered on a free electron, which is, before the scattering, at rest in the laboratory frame. After the scattering the energy of the photon is $E'_{ph}$, and the energy of the electron is $E'_e$. The photon is scattered in a direction which makes an angle $\theta$ with the direction of the incoming photon.

a) The center of mass (CM) system is an inertial reference frame, where the total momentum of the particle system vanishes. Using the transformation formulas for energy and momentum, we obtain, for the photon and electron momenta in the CM-system,

$$\bar{p}_{ph} = \frac{\bar{E}_{ph}}{c} = \gamma(p_{ph} - \beta\frac{E_{ph}}{c}) = \gamma(1-\beta)\frac{E_{ph}}{c},$$
$$\bar{p}_e = -\gamma\beta\frac{E_e}{c} = -\gamma\beta m_e c\,, \tag{8.23}$$

where $\beta$ and $\gamma$ refer to the relative velocity of the two reference systems. In the CM-system we have $\bar{p}_{ph} + \bar{p}_e = 0$, which gives

$$\gamma(1-\beta)\frac{E_{ph}}{c} = \gamma\beta m_e c \quad \Rightarrow \quad (1-\beta)E_{ph} = \beta m_e c^2\,. \tag{8.24}$$

Extracting the relative velocity of the reference systems from this we find

$$v = \frac{E_{ph}}{E_{ph} + mc^2}c = 0.164c\,, \tag{8.25}$$

where we have used $E_{ph} = 100 keV$ and $m_e c^2 = 0.51\,\mathrm{MeV}$.

b) Energy conservation gives, in the CM-system,

$$\bar{E}_{ph} + \bar{E}_e = \bar{E}'_{ph} + \bar{E}'_e\,, \tag{8.26}$$

which we rewrite as

$$\bar{E}_{ph} + \sqrt{\bar{p}_e^2 c^2 + m_e c^2} = \bar{E}'_{ph} + \sqrt{\bar{p'}_e^2 c^2 + m_e c^2}\,. \tag{8.27}$$

In the CM system we also have $\bar{p}_e^2 c^2 = \bar{p}_{ph}^2 c^2 = \bar{E}_{ph}^2$, with similar identities for the outgoing particles. This implies

$$\bar{E}_{ph} + \sqrt{\bar{E}_{ph}^2 + m_e c^2} = \bar{E}'_{ph} + \sqrt{\bar{E'}_{ph}^2 + m_e c^2}\,. \tag{8.28}$$

We note that the same function appears on both sides of the equation, and that this equation is monotonically increasing with the argument. This implies that arguments on both sides have to be equal, thus $\bar{E}_{ph} = \bar{E}'_{ph}$.

To find the value of this energy we use the Lorentz transformation formula between the reference frames $S$ and $\bar{S}$,

$$\bar{E}_{ph} = \gamma(E_{ph} - \beta p_{ph} c) = \gamma(1-\beta)E_{ph} = 84.7\,\mathrm{keV}\,. \tag{8.29}$$

c) If $\theta = 90°$ the $x$-component of $\mathbf{p}'_{ph}$ vanishes, and the transformation formula for the outgoing photon simplifies to $\bar{E}'_{ph} = \gamma E'_{ph}$, which gives

$$E'_{ph} = \frac{1}{\gamma}\bar{E}_{ph} = (1-\beta)E_{ph} = 83.6\,\text{keV}. \tag{8.30}$$

Energy conservation then determines the energy of the outgoing electron,

$$\begin{aligned} E'_e &= E_e + E_{ph} - E'_{ph} \\ &= m_e c^2 + \beta E_{ph} \\ &= 527\,\text{keV}\,. \end{aligned} \tag{8.31}$$

### *Problem 8.5*

A Lambda particle ($\Lambda$) has momentum $\bar{\mathbf{p}}_\Lambda$ along the $x$-axis in the laboratory frame $\bar{S}$. The energy of the particle in $\bar{S}$ is $\bar{E}_\Lambda = 3\,\text{GeV}(= 3000\,\text{MeV})$. The mass of $\Lambda$ is $m_\Lambda = 1116\,\text{MeV/c}^2$. In its rest frame $S$, the life time of $\Lambda$ is $\tau_\Lambda = 2.63 \times 10^{-10}s$. The $\Lambda$ particle decays to a nucleon $N$ and a pion $\pi$. They have masses $m_N = 940\,\text{MeV}$ and $m_\pi = 140\,\text{MeV}$, respectively.

a) The problem to solve here is to find the distance travelled by the Lambda particle within its life time. The gamma factor of the Lorentz transformation between the lab frame and the rest frame of $\Lambda$ is

$$\gamma = \frac{\bar{E}_\Lambda}{m_\Lambda c^2} = 2.688. \tag{8.32}$$

The life time of the particle in the lab frame, $\bar{\tau}_\Lambda$, is longer than in its rest frame, due to the relativistic time dilatation effect,

$$\bar{\tau}_\Lambda = \gamma\tau_\Lambda = 7.1 \cdot 10^{-10}\,\text{s}\,. \tag{8.33}$$

With $v$ as the velocity of $\Lambda$ in the lab frame, the distance travelled by the particle in the time $\tau_\Lambda$ is

$$d = v\bar{\tau}_\Lambda = \sqrt{1 - \frac{1}{\gamma^2}}\,c\,\bar{\tau}_\Lambda = \sqrt{\bar{\tau}_\Lambda^2 - \tau_\Lambda^2}\,c = 0.20\,\text{m}\,. \tag{8.34}$$

b) We determine next the energies of the nucleon and the pion in the rest frame $S$ of $\Lambda$. In this reference frame conservation of energy and momentum in the decay of $\Lambda$ is expressed as

$$E_N + E_\pi = m_\Lambda c^2\,, \quad \mathbf{p}_N + \mathbf{p}_\pi = 0\,, \tag{8.35}$$

and the energy-momentum relations are

$$E_N^2 = p_N^2 c^2 + m_N^2 c^4\,, \quad E_\pi^2 = p_\pi^2 c^2 + m_\pi^2 c^4\,. \tag{8.36}$$

From these equations we derive the following two equations

$$E_N^2 = (m_\Lambda c^2 - E_\pi)^2,$$
$$E_N^2 - m_N^2 c^4 = E_\pi^2 - m_\pi^2 c^4\,. \tag{8.37}$$

By combining these $E_N$ can be eliminated and $E_\pi$ determined as

$$E_\pi = \frac{m_\Lambda^2 + m_\pi^2 - m_N^2}{2m_\Lambda} c^2 = 171\,\text{MeV}\,. \tag{8.38}$$

The energy $E_N$ is then determined as

$$E_N = m_\Lambda c^2 - E_\pi = \frac{m_\Lambda^2 + m_N^2 - m_\pi^2}{2m_\Lambda} c^2 = 945\,\text{MeV}\,. \tag{8.39}$$

c) To determine the energies $\bar{E}_\pi$ and $\bar{E}_N$ in the lab frame, we need the momenta in the rest frame of $\Lambda$,

$$cp_N = cp_\pi = \sqrt{E_\pi^2 - m_\pi^2 c^2} = 98.0\,\text{MeV}\,. \tag{8.40}$$

In the Lorentz transformation formula we also need

$$\beta = \sqrt{1 - \frac{1}{\gamma^2}} = 0.928\,. \tag{8.41}$$

With the angle of the pion momentum relative to the $x$-axis given as $45°$ we have

$$(p_\pi)_x = -(p_N)_x = p_\pi/\sqrt{2}\,, \tag{8.42}$$

and the transformation to the lab frame then gives

$$\bar{E}_\pi = \gamma(E_\pi + \frac{1}{\sqrt{2}}\beta c p_\pi) = 632\,\text{MeV} \tag{8.43}$$

and

$$\bar{E}_N = \bar{E}_\Lambda - \bar{E}_\pi = 2368\,\text{MeV}\,. \tag{8.44}$$

d) To determine the angle $\bar{\theta}_\pi$ we use the transformation formula for the $y$-component of the $\pi$-momentum

$$\bar{p}_\pi \sin\bar{\theta}_\pi = p_\pi \sin\theta_\pi \quad \Rightarrow \quad \sin\bar{\theta}_\pi = \frac{1}{\sqrt{2}}\frac{p_\pi}{\bar{p}_\pi}\,. \tag{8.45}$$

By use of the energy-momentum relation $c^2\bar{p}_\pi^2 = \bar{E}_\pi^2 - m_\pi^2 c^4$ we rewrite this as

$$\sin\bar{\theta}_\pi = \frac{1}{\sqrt{2}}\frac{cp_\pi}{\sqrt{\bar{E}_\pi^2 - m_\pi^2 c^4}} = 0.112\,. \tag{8.46}$$

This corresponds to the angle $\bar{\theta}_\pi = 6.45°$. For the angle of the nucleon we similarly find

$$\sin\bar{\theta}_N = -\frac{1}{\sqrt{2}}\frac{cp_\pi}{\sqrt{\bar{E}_N^2 - m_N^2 c^4}} = -0.0319 \tag{8.47}$$

which corresponds to the angle $\bar{\theta}_N = -1.83°$.

### *Problem 8.6*

The relativistic Lagrangian is

$$L = -mc^2\sqrt{1-\frac{v^2}{c^2}} - e\Phi + e\mathbf{v}\cdot\mathbf{A}\,. \tag{8.48}$$

In order to show that the corresponding Lagrange's equation reproduces the standard, relativistic equation for a charged particle in an electromagnetic field, we first derive the partial derivatives of $L$ with respect to the particle's coordinates $x_i$ and their time derivatives,

$$\begin{aligned}\frac{\partial L}{\partial \dot{x}_i} &= \frac{m\dot{x}_i}{\sqrt{1-\frac{v^2}{c^2}}} + eA_i = \gamma m\dot{x}_i + eA_i\,,\\ \frac{\partial L}{\partial x_i} &= -e\frac{\partial \Phi}{\partial x_i} + e\mathbf{v}\cdot\frac{\partial \mathbf{A}}{\partial x_i}\,.\end{aligned} \tag{8.49}$$

Lagrange's equation then gives

$$\frac{d}{dt}(\gamma m\dot{x}_i + eA_i) + e\frac{\partial \Phi}{\partial x_i} - e\mathbf{v}\cdot\frac{\partial \mathbf{A}}{\partial x_i} = 0\,, \tag{8.50}$$

which we rewrite as

$$\frac{d}{dt}(\gamma m\dot{x}_i) + e\frac{\partial \Phi}{\partial x_i} + e\frac{\partial A_i}{\partial t} + e\dot{x}_j(\frac{\partial A_i}{\partial x_j} - \frac{\partial A_j}{\partial x_i}) = 0\,, \tag{8.51}$$

with sum over repeated index $j$. We have here used the expression for the total time derivative

$$\frac{dA_i}{dt} = \dot{x}_j\frac{\partial A_i}{\partial x_j} + \frac{\partial A_i}{\partial t}\,. \tag{8.52}$$

The contribution from the electric field we recognize as

$$eE_i = -e(\frac{\partial \Phi}{\partial x_i} + \frac{\partial A_i}{\partial t})\,. \tag{8.53}$$

To find the contribution from the magnetic field, we express the magnetic field in terms of the vector potential in the expression,

$$\begin{aligned}e(\mathbf{v}\times\mathbf{B})_i &= e\epsilon_{ijk}\dot{x}_j(\boldsymbol{\nabla}\times\mathbf{A})_k\\ &= e\epsilon_{ijk}\epsilon_{klm}x_j\partial_l A_m\\ &= e(\delta_{il}\delta_{jm} - \delta_{im}\delta_{jl})\dot{x}_j\frac{\partial A_m}{\partial x_l}\\ &= e\dot{x}_j(\frac{\partial A_j}{\partial x_i} - \frac{\partial A_i}{\partial x_j})\,.\end{aligned} \tag{8.54}$$

We find both expressions (8.53) and (8.54) in Eq. (8.51), which then can be written as

$$\frac{d}{dt}(\gamma m\dot{x}_i) = eE_i + e(\mathbf{v}\times\mathbf{B})_i\,, \tag{8.55}$$

or in vector form

$$\frac{d}{dt}(\gamma m\mathbf{v}) = e\mathbf{E} + e\mathbf{v}\times\mathbf{B}\,. \tag{8.56}$$

This is the correct relativistic form of the equation of motion for a charged particle in the electromagnetic field, where the only change compared to the non-relativistic equation is the appearance of the factor $\gamma$ in the expression for the (mechanical) momentum of the particle.

***Problem 8.7***

a) The equation of motion for a charged particle in an electric field is

$$\frac{d}{dt}\mathbf{p} = e\mathbf{E}\,. \tag{8.57}$$

When $\mathbf{E}$ is constant, and the initial condition is $\mathbf{p}(0) = 0$, the solution is

$$\mathbf{p} = e\mathbf{E}t\,. \tag{8.58}$$

The relativistic expression for the energy is

$$\mathcal{E} = \sqrt{\mathbf{p}^2c^2 + m_e^2c^4} = \gamma m_e c^2\,, \tag{8.59}$$

which gives for the gamma factor

$$\gamma = \frac{1}{m_e^2c^2}\sqrt{\mathbf{p}^2c^2 + m_e^2c^4} = \sqrt{1 + \frac{e^2E^2}{m_e^2c^4}t^2}\,. \tag{8.60}$$

This is of the form $\gamma = \sqrt{1+\kappa^2t^2}$, with $\kappa = \frac{eE}{m_ec^2}$.

b) We invert the expression found for $\gamma$, and introduce the definition $\gamma = \cosh\kappa\tau$,

$$t = \frac{1}{\kappa}\sqrt{\gamma^2-1} = \frac{1}{\kappa}\sqrt{\cosh^2\kappa\tau - 1} = \frac{1}{\kappa}\sinh\kappa\tau\,. \tag{8.61}$$

Differentiation of this with respect to $\tau$ gives

$$\frac{dt}{d\tau} = \cosh\kappa\tau = \gamma\,. \tag{8.62}$$

This shows that $\tau$ satisfies the condition to be the proper time of the electron.

c) For linear motion, the equation of motion of the electron can be written as

$$\frac{d}{dt}(\gamma m_e v) = eE\,, \tag{8.63}$$

which gives

$$\gamma m_e a + \dot{\gamma} m_e v = eE\,. \tag{8.64}$$

The time derivative of $\gamma$ is found as

$$\gamma = (1 - \frac{v^2}{c^2})^{-\frac{1}{2}} \quad \Rightarrow \quad \dot{\gamma} = (1 - \frac{v^2}{c^2})^{-\frac{3}{2}}\frac{va}{c^2} = \gamma^3\frac{va}{c^2}\,. \tag{8.65}$$

Inserted in Eq. (8.64) this gives

$$\gamma m_e a(1 + \gamma^2\frac{v^2}{c^2}) = eE \quad \Rightarrow \quad \gamma^3 a = eE/m_e. \tag{8.66}$$

Since the proper acceleration $a_0$, in the case of linear motion, is related to the acceleration $a$, measured in any inertial reference frame, by the formula $a_0 = \gamma^3 a$, the result above shows that for linear motion of a charged particle in an electric field, the proper acceleration is given by $a_0 = eE/m_e$.

### *Problem 8.8*

A particle with charge $q$ and mass $m$ moves with relativistic speed through a region $0 < x < L$ where a constant electric field $\mathbf{E}$ is directed along the $y$-axis, as indicated in the figure. The particle enters the field at $x = 0$ with momentum $\mathbf{p}_0$ in the direction orthogonal to the field. The relativistic energy at this point is denoted $\mathcal{E}_0$.

a) We determine here the time dependent momentum $\mathbf{p}(t)$, the relativistic energy $\mathcal{E}(t)$ and the gamma factor $\gamma(t)$. The equation of motion of the charged particle is

$$\frac{d\mathbf{p}}{dt} = q\mathbf{E}\,, \tag{8.67}$$

with the solution

$$\mathbf{p}(t) = \mathbf{p}_0 + q\mathbf{E}t\,. \tag{8.68}$$

Since $\mathbf{p}_0 \cdot \mathbf{E} = 0$, the energy of the particle is

$$\begin{aligned}\mathcal{E}(t) &= \sqrt{c^2p^2 + m^2c^4} \\ &= \sqrt{c^2p_0^2 + m^2c^4 + (qEct)^2} \\ &= \sqrt{\mathcal{E}_0^2 + (qEct)^2}\,. \end{aligned} \tag{8.69}$$

The relativistic gamma factor is then

$$\gamma(t) = \frac{\mathcal{E}(t)}{mc^2} = \frac{\mathcal{E}_0}{mc^2}\sqrt{1 + (\frac{qE_0}{\mathcal{E}_0}ct)^2}\,. \tag{8.70}$$

b) We next determine the components of the particle velocity $\mathbf{v} = \mathbf{p}/(\gamma m)$,

$$\begin{aligned} v_x &= \frac{p_0}{\gamma m} = \frac{p_0 c^2}{\sqrt{\mathcal{E}_0^2 + (qEct)^2}}\,, \\ v_y &= \frac{qEt}{\gamma m} = \frac{qEc^2 t}{\sqrt{\mathcal{E}_0^2 + (qEct)^2}}\,. \end{aligned} \tag{8.71}$$

There is no force acting in the $x$-direction, which implies that $p_x$ is constant. However, the velocity component $v_x$ decreases with time as a consequence of the increasing value of $\gamma$. This can be understood as a time dilatation effect, which follows from the fact that the velocity $v$ of the particle increases.

c) The derivative of the $x$-coordinate with respect to proper time $\tau$ is

$$\frac{dx}{d\tau} = \frac{dx}{dt}\frac{dt}{d\tau} = v_x\gamma = \frac{p_0}{m}\,. \tag{8.72}$$

This implies that the increase in proper time, $\Delta\tau$ during the transit of the interval $0 < x < L$, is

$$\Delta\tau = \int_0^{\Delta\tau} d\tau = \frac{m}{p_0}\int_0^L dx = \frac{m}{p_0}L\,. \tag{8.73}$$

Thus, the proper time interval $\Delta\tau$ is proportional to the length $L$ with the proportionality factor $\alpha = m/p_0$.

d) To find the transit time $\Delta t$, we change the integration variable to $t$ in the expression for $\Delta\tau$,

$$\begin{aligned} \Delta\tau &= \int_0^{\Delta t} \frac{1}{\gamma(t)}dt \\ &= \frac{mc^2}{\mathcal{E}_0}\int_0^{\Delta t} \frac{1}{\sqrt{1 + (\frac{qEct}{\mathcal{E}_0})^2}}dt \\ &= \frac{mc^2}{qEc}\int_0^{\frac{qEc}{\mathcal{E}_0}\Delta t} \frac{1}{\sqrt{1+z^2}}dz \\ &= \frac{mc^2}{qEc}\operatorname{arcsinh}(\frac{qEc}{\mathcal{E}_0}\Delta t)\,, \end{aligned} \tag{8.74}$$

which gives

$$\Delta t = \frac{\mathcal{E}_0}{qEc}\sinh(\frac{qEc}{mc^2}\Delta\tau) = \frac{\mathcal{E}_0}{qEc}\sinh(\frac{qEL}{p_0 c})\,. \tag{8.75}$$

# PART 3
# Electrodynamics

# Chapter 9

# Maxwell's equations

### *Problem 9.1*

A vector potential is expressed in cylindrical coordinates as

$$\mathbf{A}(\rho, \phi, z) = A(\phi\rho\, e^{-(\rho/a)^2}\mathbf{e}_\rho + b\, e^{-(\rho/a)^2}\mathbf{e}_z). \tag{9.1}$$

We determine first the corresponding magnetic field $\mathbf{B} = \boldsymbol{\nabla} \times \mathbf{A}$ by use of the formulas for the curl in cylindrical coordinates,

$$\begin{aligned} B_\rho &= \frac{1}{\rho}\frac{\partial A_z}{\partial \phi} - \frac{\partial A_\phi}{\partial z} = 0\,, \\ B_\phi &= \frac{\partial A_\rho}{\partial z} - \frac{\partial A_z}{\partial \rho} = -2A\frac{b}{a^2}\rho e^{-(\rho/a)^2}\,, \\ B_z &= \frac{1}{\rho}\frac{\partial(\rho A_\phi)}{\partial \rho} - \frac{1}{\rho}\frac{\partial A_\rho}{\partial \phi} = Ae^{-(\rho/a)^2}\,. \end{aligned} \tag{9.2}$$

In vector form this is

$$\mathbf{B}(\rho, \phi, z) = -2A\frac{b}{a^2}\rho e^{-(\rho/a)^2}\mathbf{e}_\phi + Ae^{-(\rho/a)^2}\mathbf{e}_z\,. \tag{9.3}$$

The current density we determine by use of Ampère's law, with $\mathbf{E} = 0$, $\mathbf{j} = \frac{1}{\mu_0}\boldsymbol{\nabla} \times \mathbf{B}$. In cylindrical coordinates the components are

$$\begin{aligned} j_\rho &= \frac{1}{\mu_0}\left[\frac{1}{\rho}\frac{\partial B_z}{\partial \phi} - \frac{\partial B_\phi}{\partial z}\right] = 0\,, \\ j_\phi &= \frac{1}{\mu_0}\left[\frac{\partial B_\rho}{\partial z} - \frac{\partial B_z}{\partial \rho}\right] = -\frac{2A}{\mu_0 a^2}\rho e^{-(\rho/a)^2}\,, \\ j_z &= \frac{1}{\mu_0}\left[\frac{1}{\rho}\frac{\partial(\rho B_\phi)}{\partial \rho} - \frac{1}{\rho}\frac{\partial B_\rho}{\partial \phi}\right] = -\frac{4Ab}{\mu_0 a^2}\left(1 - \frac{\rho^2}{a^2}\right)e^{-(\rho/a)^2}\,, \end{aligned} \tag{9.4}$$

which in vector form is expressed as

$$\mathbf{j}(\rho, \phi, z) = -\frac{2A}{\mu_0 a^2}\rho e^{-(\rho/a)^2}\mathbf{e}_\phi - \frac{4Ab}{\mu_0 a^2}\left(1 - \frac{\rho^2}{a^2}\right)e^{-(\rho/a)^2}\mathbf{e}_z\,. \tag{9.5}$$

### *Problem 9.2*

Two inertial reference frames, $S$ and $S'$, are related by the Lorentz transformation

$$t' = \gamma(t - \frac{v}{c^2}x)\,, \quad x' = \gamma(x - vt)\,, \quad y' = y\,, \quad z' = z\,. \tag{9.6}$$

A point charge $q$ sits at rest at the origin of $S$, and the electromagnetic field in this reference frame is therefore a pure Coulomb field. We will examine the electromagnetic field in $S'$.

a) The Lorentz transformations of the potentials have the general form

$$\begin{aligned} \Phi'(\mathbf{r}', t') &= \gamma(\Phi(\mathbf{r}, t) - vA_x(\mathbf{r}, t))\,, \\ A'_x(\mathbf{r}', t') &= \gamma(A_x(\mathbf{r}, t) - \frac{v}{c^2}\Phi(\mathbf{r}, t))\,, \\ A'_y(\mathbf{r}', t') &= A_y(\mathbf{r}, t)\,, \\ A'_z(\mathbf{r}', t') &= A_z(\mathbf{r}, t)\,, \end{aligned} \tag{9.7}$$

with the potentials in reference frame $S$ here given by

$$\Phi(\mathbf{r}) = \frac{q}{4\pi\epsilon_0 r}, \qquad A_x = A_y = A_z = 0\,. \tag{9.8}$$

To find the potentials expressed in the coordinates of $S'$, we make use of the coordinate transformations from $S$ to $S'$,

$$\begin{aligned} \Phi'(x', y', z', t') &= \gamma\Phi(\gamma(x' + vt'), y', z') \\ &= \frac{\gamma q}{4\pi\epsilon_0\sqrt{\gamma^2(x' + vt')^2 + y'^2 + z'^2}}\,, \\ A'_x(x', y', z', t') &= -\gamma\frac{v}{c^2}\Phi(\gamma(x' + vt'), y', z') \\ &= -\frac{\gamma\beta q}{4\pi\epsilon_0 c\sqrt{\gamma^2(x' + vt')^2 + y'^2 + z'^2}}\,, \\ A'_y &= A'_z = 0\,. \end{aligned} \tag{9.9}$$

b) The corresponding electric and magnetic fields in the reference frame $S'$ are

$$\mathbf{E}' = -\boldsymbol{\nabla}\Phi' - \frac{\partial}{\partial t'}\mathbf{A}'\,, \quad \mathbf{B}' = \boldsymbol{\nabla} \times \mathbf{A}'\,. \tag{9.10}$$

We evaluate these for each component, first for the electric field

$$E'_x = -\frac{\partial}{\partial x'}\Phi' - \frac{\partial}{\partial t'}A'_x = \frac{q}{4\pi\epsilon_0}\frac{\gamma(x'+vt')}{\left[\sqrt{\gamma^2(x'+vt')^2+y'^2+z'^2}\right]^3},$$

$$E'_y = -\frac{\partial}{\partial y'}\Phi' - = \frac{q}{4\pi\epsilon_0}\frac{\gamma y'}{\left[\sqrt{\gamma^2(x'+vt')^2+y'^2+z'^2}\right]^3},$$

$$E'_z = -\frac{\partial}{\partial z'}\Phi' - = \frac{q}{4\pi\epsilon_0}\frac{\gamma z'}{\left[\sqrt{\gamma^2(x'+vt')^2+y'^2+z'^2}\right]^3}. \tag{9.11}$$

The components of the magnetic field in moving reference frame $S'$ are

$$B'_x = 0,$$

$$B'_y = \frac{\partial}{\partial z'}A'_x = \frac{q}{4\pi\epsilon_0 c}\frac{\gamma\beta z'}{\left[\sqrt{\gamma^2(x'+vt')^2+y'^2+z'^2}\right]^3},$$

$$B'_z = -\frac{\partial}{\partial y'}A'_x = -\frac{q}{4\pi\epsilon_0 c}\frac{\gamma\beta y'}{\left[\sqrt{\gamma^2(x'+vt')^2+y'^2+z'^2}\right]^3}. \tag{9.12}$$

c) We now restrict the scalar potential to the two-dimensional $x', y'$-plane at time $t' = 0$,

$$\Phi'(x', y') = \frac{\gamma q}{4\pi\epsilon_0\sqrt{\gamma^2 x'^2 + y'^2}}, \tag{9.13}$$

and similarly the electric field

$$E'_x(x', y') = \frac{q}{4\pi\epsilon_0}\frac{\gamma x'}{\left[\sqrt{\gamma^2 x'^2 + y'^2}\right]^3},$$

$$E'_y(x', y') = \frac{q}{4\pi\epsilon_0}\frac{\gamma y'}{\left[\sqrt{\gamma^2 x'^2 + y'^2}\right]^3}. \tag{9.14}$$

Since $E'_z$ is proportional to $z'$, it vanishes in the $x', y'$-plane.

A contour plot of the scalar potential $\Phi'$ and field lines of the electric field $\mathbf{E}'$, in reference system $S'$, are shown in the left panel of Fig. 9.1. There, we have used the value $\gamma = 5/3$. For comparison the corresponding diagram, with potential $\Phi$ and field lines of $\mathbf{E}$ in reference $S$ are shown in the right panel. The equipotential curves of $\Phi'$ are seen to be squeezed in the $x'$-direction, compared to the rotationally symmetric curves of $\Phi$. The field lines of $\mathbf{E}'$ are, on the other hand, radially directed, in the same way

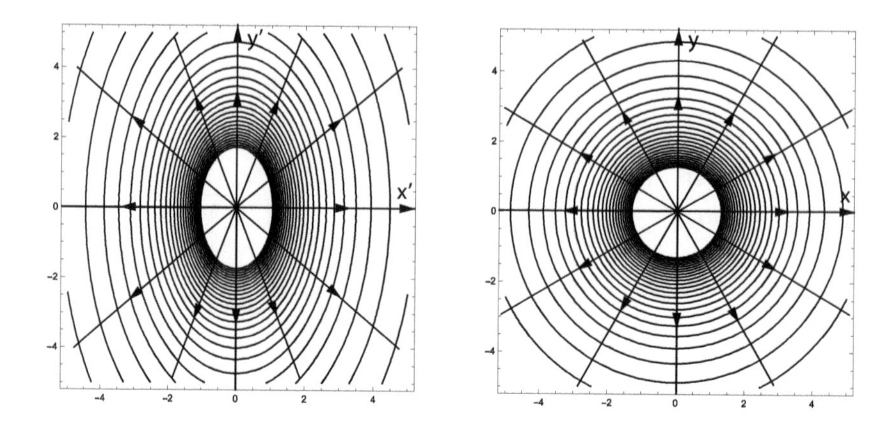


Fig. 9.1 The scalar potential and the electric field lines of a point charge. In the left panel the charge is moving with velocity in the $x$-direction, and in the right panel the charge is at rest.

as the field lines of $\mathbf{E}$. This is a consequence of the ratio between the vector components of the field being $\mathbf{E}'_x/\mathbf{E}'_y = x'/y'$. However the absolute value of the field $\mathbf{E}'$ is not rotationally invariant, as shown by the expression

$$|\mathbf{E}'(x',y')| = \frac{\gamma q}{4\pi\epsilon_0}\frac{\sqrt{x'^2+y'^2}}{\left[\sqrt{\gamma^2 x'^2+y'^2}\right]^3}. \tag{9.15}$$

In the figure this effect appears in the form of a denser set of field lines in the $y$-direction than in the $x$-direction.

### *Problem 9.3*

Here we will show how the following higher order Lorentz invariants of the electromagnetic field,

$$I_3 \equiv F^{\mu\nu}F_{\nu\lambda}F^{\lambda}{}_{\mu}\,, \quad I_4 \equiv F^{\mu\nu}F_{\nu\lambda}F^{\lambda\rho}F_{\rho\mu}\,, \tag{9.16}$$

can be expressed in terms of the quadratic invariants

$$I_1 = \frac{1}{2}F^{\mu\nu}F_{\mu\nu}\,, \quad I_2 = \frac{1}{4}F^{\mu\nu}\tilde{F}_{\mu\nu}\,. \tag{9.17}$$

a) By making use of the antisymmetry of $F^{\mu\nu}$, and swopping names of summation indices, one can change the sign of $I_3$,

$$\begin{aligned} I_3 &= F^{\mu\nu}F_{\nu\lambda}F^{\lambda}{}_{\mu} = F^{\nu\mu}F_{\lambda\nu}F^{\lambda}{}_{\mu} \\ &= F^{\mu\nu}F^{\lambda}{}_{\nu}F_{\lambda\mu} = F^{\mu\nu}F_{\lambda\nu}F^{\lambda}{}_{\mu} \\ &= -F^{\mu\nu}F_{\nu\lambda}F^{\lambda}{}_{\mu}\,. \end{aligned} \tag{9.18}$$

This shows that $I_3 = -I_3$ which implies $I_3 = 0$.

b) To study $I_4$ we choose coordinate axes which simplify the fields in the following way,

$$\mathbf{B} = B\mathbf{i}\,, \quad \mathbf{E} = E_1\mathbf{i} + E_2\mathbf{j}\,. \tag{9.19}$$

This implies that the field tensor has only six non-vanishing matrix elements, $F^{01}$, $F^{02}$, $F^{23}$ and their transposed. The invariant $I_4$, expressed as a sum over products of non-vanishing matrix elements, is

$$\begin{aligned}
I_4 &= (F^{01}F_{10}F^{01}F_{10} + F^{10}F_{01}F^{10}F_{01}) \\
&= (F^{01}F_{10}F^{02}F_{20} + F^{10}F_{02}F^{20}F_{01}) \\
&= (F^{02}F_{20}F^{01}F_{10} + F^{20}F_{01}F^{10}F_{02}) \\
&= (F^{02}F_{20}F^{02}F_{20} + F^{20}F_{02}F^{20}F_{02}) \\
&= (F^{02}F_{23}F^{32}F_{20} + F^{20}F_{02}F^{23}F_{32}) \\
&= (F^{23}F_{32}F^{20}F_{02} + F^{32}F_{20}F^{02}F_{23}) \\
&= (F^{23}F_{32}F^{23}F_{32} + F^{32}F_{23}F^{32}F_{23}) \\
&= \frac{2}{c^4}(E_1^4 + 2E_1^2E_2^2 + E_2^4) - \frac{4}{c^2}E_2^2B^2 + 2B^4\,.
\end{aligned} \tag{9.20}$$

The invariants $I_1$ and $I_2$ can be written as

$$I_1 = B^2 - \frac{1}{c^2}(E_1^2 + E_2^2)\,, \quad I_2 = \frac{1}{c}E_1 B\,, \tag{9.21}$$

and by comparing the above expressions we find that they produce the identity

$$I_4 = 2I_1^2 + 4I_2^2\,. \tag{9.22}$$

## *Problem 9.4*

We study the bending of a proton's trajectory in one of the magnets at the accelerator ring LHC at CERN. We have the following information about the proton trajectory. The proton momentum is $p = 7.0\,\mathrm{TeV/c}$, the bending radius of the magnet is $R = 2804\,\mathrm{m}$, and the strength of the magnetic field is $B = 8.33\,\mathrm{T}$. The proton mass is $m = 938\,\mathrm{MeV/c^2}$.

a) The equation of motion in the magnetic field is

$$\dot{\mathbf{p}} = e\mathbf{v}\times\mathbf{B}\,, \quad \mathbf{p} = \gamma m\mathbf{v}\,. \tag{9.23}$$

This shows that $\dot{\mathbf{p}}\cdot\mathbf{v} = 0$ which implies that $v$ and therefore $\gamma$ are constants of motion. The bending radius of the electron orbit is related to the particle's acceleration by $a = v^2/R$, with $a$. In the present case we have $\gamma ma = evB$, which gives

$$\gamma m \frac{v^2}{R} = evB \quad \Rightarrow \quad eBR = \gamma mv = p\,. \tag{9.24}$$

To check this numerically it is convenient to write the relation as $eBRc = pc$, where the two terms then have the dimension of energy. With $B$, $c$, and $R$ given in SI-units, the unit charge $e$ will effectively change the energy unit of the product $eBRc$ to electron volt. We insert the numerical values in the left-hand side of the equation above,

$$eBRc = 8.33 \cdot (3.0 \cdot 10^8) \cdot 2804\,\text{eV} = 7.01 \cdot 10^{12}\,\text{eV}\,, \tag{9.25}$$

which fits well the given value for the proton momentum.

b) The relativistic energy-momentum relation gives

$$E^2 = p^2c^2 + m^2c^4 = \gamma^2 m^2 c^4 \quad \Rightarrow \quad \gamma^2 = \frac{p^2}{m^2c^2} + 1\,, \tag{9.26}$$

which determines $\gamma$ as

$$\gamma = \sqrt{\frac{p^2}{m^2c^2} + 1} \approx \frac{p}{mc} = 7463\,. \tag{9.27}$$

The acceleration is

$$a = \frac{v^2}{R} \approx \frac{c^2}{R} = 3.2 \cdot 10^{13} m/s^2\,. \tag{9.28}$$

c) We consider the same situation in the instantaneous rest frame of the proton. Since $\mathbf{E} = 0$ and $\mathbf{v} \cdot \mathbf{B} = 0$, the general transformation formula for the fields, is here simplified to

$$\mathbf{B}' = \gamma \mathbf{B}\,; \quad \mathbf{E}' = \gamma \mathbf{v} \times \mathbf{B}\,, \tag{9.29}$$

with $\mathbf{E}'$ and $\mathbf{B}'$ as the fields in the instantaneous inertial rest frame of a proton in the accelerator ring. This gives for the magnetic field

$$B' = \gamma B = 62167\,\text{T}\,. \tag{9.30}$$

This is, like the magnetic field in the rest frame of the ring, directed orthogonally to the plane of the ring. The electric field strength is

$$\begin{aligned} E' &= \gamma v B = vB' \\ &= 3.0 \cdot 10^8 \cdot 62167\,\text{Tm/s} \\ &= 1.86 \cdot 10^{13}\,\text{V/m}\,. \end{aligned} \tag{9.31}$$

The proper acceleration, which is the acceleration in the instantaneous inertial reference frame, is

$$\mathbf{a}_0 = \frac{e}{m}\mathbf{E}' = \gamma\frac{e}{m}\mathbf{v}\times\mathbf{B} = \gamma^2\mathbf{a}\,, \tag{9.32}$$

with $\mathbf{a}$ as the acceleration in the lab frame. The numerical value is

$$a_0 = 7463^2 \cdot 3.2 \cdot 10^{13}\,\mathrm{m/s^2} = 1.78 \cdot 10^{21}\,\mathrm{m/s^2}\,. \tag{9.33}$$

## Chapter 10

# Electromagnetic field dynamics

*Problem 10.1*

We consider the following case of the electromagnetic wave, where the **E** and **B** fields are parallel,

$$\begin{aligned}\mathbf{E}(\mathbf{r},t) &= E_0 \sin\omega t(\sin kx\,\mathbf{e}_y + \cos kx\,\mathbf{e}_z)\,,\\ \mathbf{B}(\mathbf{r},t) &= \frac{1}{c}E_0 \cos\omega t(\sin kx\,\mathbf{e}_y + \cos kx\,\mathbf{e}_z)\,.\end{aligned} \tag{10.1}$$

a) Since the **E** and **B** fields factorize in a time dependent and a space dependent part, this implies that the fields at different positions **r** oscillate in phase. This means that the wave is non-propagating, it is a standing wave.

We next check that the given electromagnetic wave satisfies the four source-free Maxwell's equations. We write

$$\begin{aligned}\mathbf{E}(x,t) &= E_y(x,t)\mathbf{e}_y + E_z(x,t)\mathbf{e}_z\,,\\ \mathbf{B}(x,t) &= B_y(x,t)\mathbf{e}_y + B_z(x,t)\mathbf{e}_z\,,\end{aligned} \tag{10.2}$$

with

$$\begin{aligned}E_y &= E_0 \sin\omega t\,\sin kx, \quad E_z = E_0 \sin\omega t\,\cos kx\,,\\ B_y &= \frac{1}{c}E_0 \cos\omega t\,\sin kx, \quad B_z = \frac{1}{c}E_0\cos\omega t\,\cos kx\,.\end{aligned} \tag{10.3}$$

I: Gauss' law;

$$\boldsymbol{\nabla}\cdot\mathbf{E} = \frac{\partial E_y}{\partial y} + \frac{\partial E_z}{\partial z} = 0. \tag{10.4}$$

This is satisfied, since there is no dependence on the coordinates $y$ and $z$.

II: The curl of **B** is

$$\begin{aligned}\boldsymbol{\nabla}\times\mathbf{B} &= \boldsymbol{\nabla}B_y \times \mathbf{e}_y + \boldsymbol{\nabla}B_z\times\mathbf{e}_z\\ &= \frac{\partial B_y}{\partial x}\mathbf{e}_x\times\mathbf{e}_y + \frac{\partial B_z}{\partial x}\mathbf{e}_x\times\mathbf{e}_z\\ &= \frac{k}{c}E_0[\cos\omega t\,\sin kx\mathbf{e}_y + \cos\omega t\,\cos kx\mathbf{e}_y]\,,\end{aligned} \tag{10.5}$$

and the time derivative of $\mathbf{E}$ is

$$\frac{\partial \mathbf{E}}{\partial t} = \omega E_0[\cos\omega t\,\sin kx\mathbf{e}_y + \cos\omega t\,\cos kx\mathbf{e}_y] = c^2\boldsymbol{\nabla}\times\mathbf{B}\,. \quad (10.6)$$

This shows that Ampère's law (without sources) is satisfied,

$$\boldsymbol{\nabla}\times\mathbf{B} - \frac{1}{c^2}\frac{\partial \mathbf{E}}{\partial t} = 0\,. \quad (10.7)$$

III: Gauss' law for the magnetic field is satisfied by the same argument as for the electric field,

$$\boldsymbol{\nabla}\cdot\mathbf{B} = \frac{\partial B_y}{\partial y} + \frac{\partial B_z}{\partial z} = 0\,. \quad (10.8)$$

IV: Faraday's law is shown in the same way as for Ampère's law. The curl of $\mathbf{E}$ is

$$\begin{aligned}\boldsymbol{\nabla}\times\mathbf{E} &= \frac{\partial E_y}{\partial x}\mathbf{e}_x\times\mathbf{e}_y + \frac{\partial E_z}{\partial x}\mathbf{e}_x\times\mathbf{e}_z \\ &= \frac{k}{c}E_0[\cos\omega t\,\sin kx\,\mathbf{e}_y + \cos\omega t\,\cos kx\,\mathbf{e}_y]\,,\end{aligned} \quad (10.9)$$

and the time derivative of $\mathbf{B}$ is

$$\frac{\partial \mathbf{B}}{\partial t} = -\frac{\omega}{c}E_0[\sin\omega t\,\sin kx\mathbf{e}_y - \sin\omega t\,\cos kx\mathbf{e}_y]\,. \quad (10.10)$$

This gives Faraday's law

$$\boldsymbol{\nabla}\times\mathbf{E} + \frac{\partial \mathbf{B}}{\partial t} = 0\,. \quad (10.11)$$

Thus, all Maxwell's equations are satisfied.

b) The momentum density is proportional to the Poynting vector

$$\mathbf{g} = \epsilon_0\mathbf{E}\times\mathbf{B} = 0\,, \quad (10.12)$$

where $\mathbf{g}$ vanishes since $\mathbf{E}$ and $\mathbf{B}$ are parallel. This can be understood as being a consequence of the fact that the wave is non-propagating. The energy density is

$$\begin{aligned}u &= \frac{1}{2}(\epsilon_0 E^2 + \frac{1}{\mu_0}B^2) \\ &= \frac{1}{2}[\epsilon_0(E_y^2 + E_z^2) + \frac{1}{\mu_0}(B_y^2 + B_z^2)] \\ &= \frac{1}{2}\epsilon_0 E_0^2[\sin^2\omega t\,(\sin^2 kx + \cos^2 kx) + \cos^2\omega t\,(\sin^2 kx + \cos^2 kx)] \\ &= \frac{1}{2}\epsilon_0 E_0^2\,.\end{aligned} \quad (10.13)$$

c) We focus first on the electric field,

$$\mathbf{E} = E_0(\sin\omega t \sin kx\, \mathbf{e}_y + \sin\omega t \cos kx\, \mathbf{e}_z)\,, \tag{10.14}$$

and make use of the following identities,

$$\begin{aligned}
\sin\omega t \sin kx &= \frac{1}{2}[\cos(kx - \omega t) - \cos(kx + \omega t)]\,, \\
\sin\omega t \cos kx &= \frac{1}{2}[-\sin(kx - \omega t) + \sin(kx + \omega t)]\,.
\end{aligned} \tag{10.15}$$

This gives

$$\begin{aligned}
\mathbf{E} &= \frac{1}{2}E_0[\cos(kx - \omega t)\mathbf{e}_y - \sin(kx - \omega t)\mathbf{e}_z] \\
&\quad -\frac{1}{2}E_0[\cos(kx + \omega t)\mathbf{e}_y - \sin(kx + \omega t)\mathbf{e}_z] \\
&\equiv \mathbf{E}_+ + \mathbf{E}_-\,.
\end{aligned} \tag{10.16}$$

The magnetic field can be decomposed in the same way, which gives

$$\begin{aligned}
\mathbf{B} &= \frac{1}{2c}E_0[\sin(kx - \omega t)\mathbf{e}_y + \cos(kx - \omega t)\mathbf{e}_z] \\
&\quad +\frac{1}{2c}E_0[\sin(kx + \omega t)\mathbf{e}_y + \cos(kx + \omega t)\mathbf{e}_z] \\
&\equiv \mathbf{B}_+ + \mathbf{B}_-\,.
\end{aligned} \tag{10.17}$$

The expressions above show that $\mathbf{E}_+$ and $\mathbf{B}_+$ define the electric and magnetic components of a a right-moving (positive $x$-direction) wave, and that $\mathbf{E}_-$ and $\mathbf{B}_-$ describe a left-moving wave. The expressions also show that each of the two waves satisfy the orthogonality conditions $\mathbf{E}_+ \cdot \mathbf{B}_+ = \mathbf{E}_- \cdot \mathbf{B}_- = 0$.

The two vector components of $\mathbf{E}_+$, and of $\mathbf{B}_+$, in the $y$- and $z$- directions are 90° out of phase, which shows that the polarization is circular. By looking at the time dependence of the rotating vectors for $x = 0$, it is straight forward to find that the rotation frequency in the $y, z$-plane is positive. This means that the polarization is right-handed, circular. For the components of $\mathbf{E}_-$, and of $\mathbf{B}_-$ we similarly find circular motion, but in the opposite direction in the $y, z$-plane. However, since the direction of propagation is also inverted, it means that in this case the polarization is also right-handed, circular.

d) It is clear from the expressions above that the right-moving as well as the left-moving waves can each be decomposed in two waves, where the electric field oscillates in either the $y$-direction or in the $z$-direction. A plane polarized wave is characterized by **E** and **B** oscillating in phase and being orthogonal. In the present case this means that the right-moving and left moving waves are decomposed in plane polarized waves in the following way,

$$\begin{aligned}
&\text{Right}-\text{moving}:\\
&\mathbf{E}_{+1}=\frac{1}{2}E_0\cos(kx-\omega t)\mathbf{e}_y\,,\quad \mathbf{B}_{+1}=\frac{1}{2c}E_0\cos(kx-\omega t)\mathbf{e}_z\,,\\
&\mathbf{E}_{+2}=-\frac{1}{2}E_0\sin(kx-\omega t)\mathbf{e}_z\,,\quad \mathbf{B}_{+2}=\frac{1}{2c}E_0\sin(kx-\omega t)\mathbf{e}_y\,,\\
&\text{Left}-\text{moving}:\\
&\mathbf{E}_{-1}=-\frac{1}{2}E_0\cos(kx+\omega t)\mathbf{e}_y\,,\quad \mathbf{B}_{-1}=\frac{1}{2c}E_0\cos(kx+\omega t)\mathbf{e}_z\,,\\
&\mathbf{E}_{-2}=\frac{1}{2}E_0\sin(kx+\omega t)\mathbf{e}_z\,,\quad \mathbf{B}_{-2}=\frac{1}{2c}E_0\sin(kx+\omega t)\mathbf{e}_y\,.
\end{aligned} \tag{10.18}$$

### *Problem 10.2*

A monochromatic plane wave of light is sent through a birefringent crystal in the $z$-direction. The wave can be decomposed in linearly polarized components, where polarization in the $x$-direction corresponds to a wave with a faster propagation velocity, $c_f$, than the velocity $c_s$, of a wave which is polarized in the $y$-direction. Inside the crystal (with $z > 0$) the wave has the form

$$\mathbf{E}(z,t)=\frac{E_0}{\sqrt{2}}[\cos(k_f z-\omega t)\mathbf{e}_x+\cos(k_s z-\omega t)\mathbf{e}_y]\,, \tag{10.19}$$

where $k_f = \omega/c_f$, $k_s = \omega/c_s$, and $E_0$ is the amplitude of the oscillating field.

a) For $z = 0$ the electric field vector is

$$\mathbf{E}(0,t)=E_0\cos\omega t\,\frac{1}{\sqrt{2}}(\mathbf{e}_x+\mathbf{e}_y)\,. \tag{10.20}$$

The expression shows that **E** oscillates in a fixed direction, determined by the unit vector $(\mathbf{e}_1+\mathbf{e}_2)/\sqrt{2}$. This means that the polarization is linear, halfway between the $x$- and the $y$-axis.

b) We introduce the following variables, $k_0 = (k_f + k_s)/2$ and $\Delta k = k_s - k_f$. This gives

$$\cos(k_f z - \omega t) = \cos[(k_0 - \frac{1}{2}\Delta k)z - \omega t)]$$
$$= \cos[k_0 z - \omega t)]\cos[\frac{1}{2}\Delta k\, z] + \sin[k_0 z - \omega t)]\sin[\frac{1}{2}\Delta k\, z]\,, \tag{10.21}$$

and

$$\cos(k_s z - \omega t) = \cos[(k_0 + \frac{1}{2}\Delta k)z - \omega t)]$$
$$= \cos[k_0 z - \omega t)]\cos[\frac{1}{2}\Delta k\, z] - \sin[k_0 z - \omega t)]\sin[\frac{1}{2}\Delta k]\,. \tag{10.22}$$

With the new variables the electric field gets the form

$$\mathbf{E}(z,t) = \frac{E_0}{\sqrt{2}}[\cos[k_0 z - \omega t)]\cos[\frac{1}{2}\Delta k\, z](\mathbf{e}_x + \mathbf{e}_y)$$
$$+ \sin[k_0 z - \omega t)]\sin[\frac{1}{2}\Delta k\, z](\mathbf{e}_x - \mathbf{e}_y)] \tag{10.23}$$

and can be rewritten as

$$\mathbf{E}(z,t) = E_{10}(z)\cos(k_0 z - \omega t)\mathbf{e}_1 + E_{20}(z)\sin(k_0 z - \omega t)\mathbf{e}_2$$
$$\equiv E_1(z,t)\mathbf{e}_1 + E_2(z,t)\mathbf{e}_2 \tag{10.24}$$

where the amplitudes are

$$E_{10}(z) = E_0\cos[\frac{1}{2}\Delta k\, z]\,, \quad E_{20}(z) = E_0\sin[\frac{1}{2}\Delta k\, z]\,, \tag{10.25}$$

and the unit vectors

$$\mathbf{e}_1 = (\mathbf{e}_x + \mathbf{e}_y)/\sqrt{2}\,, \quad \mathbf{e}_2 = (\mathbf{e}_x - \mathbf{e}_y)/\sqrt{2}\,. \tag{10.26}$$

c) The amplitudes satisfy the equation

$$\frac{E_1^2}{E_{10}^2} + \frac{E_2^2}{E_{20}^2} = 1\,. \tag{10.27}$$

For fixed $z$, with variable $t$, this describes an ellipse, which shows that the polarization is elliptic, with symmetry axes in the directions of $\mathbf{e}_1$ and $\mathbf{e}_2$. The eccentricity is determined by the ratio $|E_{10}|/|E_{20}|$. This ratio changes continuously as $z$ changes, so the polarization changes continuously through the crystal from linear to circular and back to linear polarization.

For $z = 0$ we have $E_{20} = 0$, which means that it is linearly polarized in the direction of $\mathbf{e}_1$, for $z = \pi/(2\Delta k)$ we have $E_{10} = E_{20}$ which means left-handed, circular polarization, and for $z = \pi/\Delta k$ there is linear polarization in the direction of $\mathbf{e}_2$. Then there is right-handed circular polarization for $z = 3\pi/(2\Delta k)$, and the wave is back to linear polarization in the direction $\mathbf{e}_1$ for $z = 2\pi/\Delta k$.

### *Problem 10.3*

A point charge $q$ is placed in a constant magnetic field $\mathbf{B}$, with the electric and magnetic fields given by

$$\mathbf{E} = \frac{q}{4\pi\epsilon_0 r^2}\mathbf{e}_r\,, \qquad \mathbf{B} = B\mathbf{k}\,. \tag{10.28}$$

Poynting's vector is

$$\mathbf{S} = \frac{1}{\mu_0}\mathbf{E}\times\mathbf{B} = \frac{qB}{4\pi\epsilon_0 r^2}\mathbf{e}_r\times\mathbf{k} = \frac{qB}{4\pi\epsilon_0 r^2}\sin\theta\,\mathbf{e}_\phi\,, \tag{10.29}$$

where $\theta$ is the angle between $\mathbf{r}$ and $\mathbf{k}$, and $\mathbf{e}_\phi$ is the angular unit vector in the $x, y$-plane. This shows that Poynting's vector, which describes the energy current density, circulates around the $z$-axis. The field momentum density, which is proportional to Poynting's vector, $\mathbf{g} = \mathbf{S}/c^2$, shows the same behavior.

## Chapter 11

# Maxwell's equations with stationary sources

### *Problem 11.1*

Three point charges, two with charge $+q$ and one with charge $-q$ are positioned in the sequence $(+q, -q, +q)$ along the $x$-axis. The distance between neighboring charges are equal to $d$, and the middle charge is placed at the origin $x = 0$.

a) The potential of the three charges is

$$\Phi(\mathbf{r}) = \frac{q}{4\pi\epsilon_0}\left(\frac{1}{|\mathbf{r} + d\,\mathbf{i}|} - \frac{1}{r} + \frac{1}{|\mathbf{r} - d\,\mathbf{i}|}\right), \tag{11.1}$$

with $\mathbf{i}$ as the unit vector along the $x$-axis. A contour plot of the potential is shown in panel (a) of Fig. 11.1.

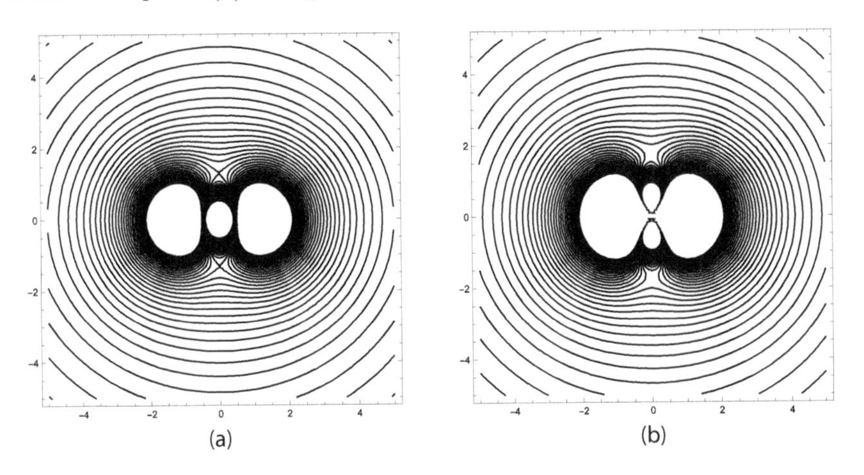


Fig. 11.1 Contour plot of the potential from three point charges $+q$, $-q$, and $+q$, located along a line. To the left the full Coulomb potential is shown, to the right the sum of contributions from the monopole and quadrupole terms in the multipole expansion of the potential.

b) The charge distribution is given by

$$\rho(\mathbf{r}) = \sum_i q_i \delta(\mathbf{r} - \mathbf{r}_i)\,, \tag{11.2}$$

with the total charge given as

$$q_{tot} = \sum_i q_i = q. \tag{11.3}$$

The electric dipole moment is

$$\mathbf{p} = \int \rho(\mathbf{r}) dV = \sum_i q_i \mathbf{r}_i = q(d - d)\mathbf{i} = 0 \tag{11.4}$$

and the quadrupole moment about the axis $\mathbf{n} = \mathbf{r}/r$ is

$$\begin{aligned} Q_{\mathbf{n}} &= \int [3(\mathbf{n} \cdot \mathbf{r}')^2 - r'^2] dV' \\ &= \sum_i q_i [3(\mathbf{n} \cdot \mathbf{r}_i)^2 - r_i^2] \\ &= 2qd^2 \left[ 3 \left( \frac{x}{r} \right)^2 - 1 \right] . \end{aligned} \tag{11.5}$$

c) The sum of the contributions from these to the scalar potential is

$$\Phi'(\mathbf{r}) = \frac{q}{4\pi\epsilon_0} \left\{ \frac{1}{r} + \frac{d^2}{r^3} \left[ 3 \left( \frac{x}{r} \right)^2 - 1 \right] \right\} . \tag{11.6}$$

A contour plot of the function is shown in panel (b) of the figure. Comparison with the plot of the full potential $\Phi(\mathbf{r})$ in panel (a) shows that the contributions from the first three terms in the multipole expansion reproduce the full potential well for distances $r$ larger than the distance between the charges, $d$, but differences appear for smaller values of $r$.

### *Problem 11.2*

A non-relativistic particle, with electric charge $q$ and mass $m$, moves in a magnetic dipole field, given by the vector potential

$$\mathbf{A} = \frac{\mu_0}{4\pi r^3} (\boldsymbol{\mu} \times \mathbf{r})\,, \tag{11.7}$$

with $\boldsymbol{\mu}$ the magnetic dipole moment of a static charge distribution centered at the origin.

a) The Lagrangian of the particle is

$$L = \frac{1}{2} m v^2 + q \mathbf{v} \cdot \mathbf{A} = \frac{1}{2} m \mathbf{v}^2 + \frac{q\mu_0}{4\pi r^3} \mathbf{v} \cdot (\boldsymbol{\mu} \times \mathbf{r})\,. \tag{11.8}$$

By use of the cyclic property of the triple product, we re-write the last term,

$$L = \frac{1}{2} m \mathbf{v}^2 + \frac{q\mu_0}{4\pi r^3} \boldsymbol{\mu} \cdot (\mathbf{r} \times \mathbf{v}) = \frac{1}{2} m \mathbf{v}^2 + \frac{q\mu_0}{4\pi m r^3} \boldsymbol{\mu} \cdot \boldsymbol{\ell} . \tag{11.9}$$

b) With the magnetic dipole moment oriented along the $z$-axis, and the particle moving in the $x, y$-plane we have

$$\boldsymbol{\mu} \cdot \boldsymbol{\ell} = |\boldsymbol{\mu}| \ell_z = |\boldsymbol{\mu}| m r^2 \dot{\phi} . \tag{11.10}$$

This gives the following expression for the Lagrangian,

$$L = \frac{1}{2} m (\dot{r}^2 + r^2 \dot{\phi}^2) + \lambda \frac{\dot{\phi}}{r} , \tag{11.11}$$

with $\lambda = q\mu_0 |\boldsymbol{\mu}| / 4\pi$. The canonical momentum $p_\phi$ conjugate to $\phi$ is

$$p_\phi = \frac{\partial L}{\partial \dot{\phi}} = m r^2 \dot{\phi} + \frac{\lambda}{r} = \ell_z + \frac{\lambda}{r} . \tag{11.12}$$

It can be interpreted as the total conserved angular momentum about the $z$-axis. The first term is the mechanical contribution from the moving particle, while the second term is an electromagnetic field contribution.

Since $L$ has no explicit time dependence the Hamiltonian is a constant of motion,

$$H = p_r \dot{r} + p_\phi \dot{\phi} - L = m \dot{r}^2 + m r^2 \dot{\phi}^2 - L = \frac{1}{2} m (\dot{r}^2 + r^2 \dot{\phi}^2) , \tag{11.13}$$

which is the conserved kinetic energy of the particle.

c) Lagrange's equation for the coordinate $r$ is

$$\frac{d}{dt} \frac{\partial L}{\partial \dot{r}} - \frac{\partial L}{\partial r} = m \ddot{r} - m r \dot{\phi}^2 + \lambda \frac{\dot{\phi}}{r^2} = 0 . \tag{11.14}$$

To eliminate $\dot{\phi}$ we write it as

$$\dot{\phi} = \frac{p_\phi}{m r^2} - \frac{\lambda}{m r^3} , \tag{11.15}$$

and exploit that $p_\phi$ is a constant. Inserted in Lagrange's equation this gives

$$m \ddot{r} - m r \left( \frac{p_\phi^2}{m^2 r^4} - 2 \frac{\lambda p_\phi}{m^2 r^5} + \frac{\lambda^2}{m^2 r^6} \right) + \frac{\lambda}{r^2} \left( \frac{p_\phi}{m r^2} - \frac{\lambda}{m r^3} \right) = 0$$

$$\Rightarrow \qquad m \ddot{r} - \frac{p_\phi^2}{m r^3} + 3 \frac{\lambda p_\phi}{m r^4} - 2 \frac{\lambda^2}{m r^5} = 0 . \tag{11.16}$$

By multiplying with $\dot{r}$, the equation can be integrated, to give

$$\frac{1}{2}m\dot{r}^2 + \frac{1}{2}\frac{p_\phi^2}{mr^2} - \frac{p_\phi \lambda}{mr^3} + \frac{1}{2}\frac{\lambda^2}{mr^4} = const.\,, \tag{11.17}$$

and with

$$\dot{\phi}^2 = \frac{p_\phi^2}{m^2 r^4} - 2\frac{p_\phi \lambda}{m^2 r^5} + \frac{\lambda^2}{m^2 r^6}\,, \tag{11.18}$$

we recognize that the expression in (11.17) is in fact the kinetic energy,

$$\frac{1}{2}m\dot{r}^2 + \frac{1}{2}mr^2\dot{\phi}^2 = T\,. \tag{11.19}$$

### *Problem 11.3*

A current $I$ is running in a rectangular wire loop ABCD, with length $a$ in the $x$-direction and length $b$ in the $y$-direction. In the rest frame $S$ of the loop, the wire is charge neutral. In a second inertial reference frame $S'$ the current loop moves with velocity $v$ in the positive $x$-direction. We use, in the following, $\mathbf{e}_x$, $\mathbf{e}_y$ and $\mathbf{e}_z$ as unit vectors along the $x$, $y$ and $z$ axes, respectively, and introduce the vectors $\mathbf{a} = a\,\mathbf{e}_x$ and $\mathbf{b} = b\,\mathbf{e}_y$.

a) Since the charge density is zero, the electric dipole moment vanishes in the rest frame of the loop. The magnetic moment is

$$\begin{aligned}
\mathbf{m} &= \frac{1}{2}\int (\mathbf{r}\times\mathbf{j}(\mathbf{r}))d^3r = \frac{1}{2}I\oint_{ABCD} \mathbf{r}\times d\mathbf{r} \\
&= \frac{1}{2}I\left[\int_0^a dx\, x\,\mathbf{e}_x \times \mathbf{e}_x + \int_0^b dy(a\,\mathbf{e}_x + y\,\mathbf{e}_y)\times\mathbf{e}_y \right.\\
&\quad \left. + \int_a^0 dx\,(x\,\mathbf{e}_x + b\,\mathbf{e}_y)\times\mathbf{e}_x + \int_b^0 dy\,y\,\mathbf{e}_y\times\mathbf{e}_y\right] \\
&= I\mathbf{a}\times\mathbf{b}
\end{aligned} \tag{11.20}$$

where at the first step the volume integral is reduced to a line integral by introducing the current $I$ as the integral of the current density $\mathbf{j}$ over the cross section of the wire.

b) Due to the motion of the current loop, when viewed in reference frame $S'$, there is a length contraction by the factor $1/\gamma$ in the $x$-direction, but no contraction in the $y$-direction. Therefore, in the $S'$ frame, the width of the rectangle in the $x$-direction is $a' = a/\gamma$, while the width in the $y$-direction is $b' = b$.

c) The Lorentz transformation formulas for the charge and current densities in the present case are

$$\rho' = \gamma(\rho + \frac{v}{c^2}j_x) = \gamma\frac{v}{c^2}j_x\,, \quad j_x' = \gamma(j_x + v\rho) = \gamma j_x\,, \tag{11.21}$$

and $j'_y = j_y$. This shows that in reference frame $S'$ the charge density vanishes on parts BC and DA of the current loop, where the current runs orthogonal to the $x$-axis. The integrated charges on the two other parts, where the current runs parallel to the $x$-axis, are

$$Q_\pm = \pm a'\gamma \frac{v}{c^2} j\Delta = \pm aI\frac{v}{c^2}\,, \tag{11.22}$$

where the current is written as $I = j\Delta$, with $\Delta$ as the cross section of the current wire. $Q_+$ represents the charge on part AB and $Q_-$ on part CD, respectively.

d) The electric dipole moment in $S'$ now gets contributions from parts AB and CD of the loop, where charge densities are $\rho_\pm = Q_\pm/(a'\Delta) = \gamma I v/(c^2\Delta)$,

$$\begin{aligned}\mathbf{p}' &= \gamma\frac{Iv}{c^2\Delta}\left[\Delta\int_0^{a'} dx\, x\,\mathbf{e}_x - \Delta\int_0^{a'} dx(x\,\mathbf{e}_x + b\,\mathbf{e}_y)\right]\\ &= -\gamma\frac{Iv}{c^2}a'b\,\mathbf{e}_y = -\frac{Iv}{c^2}ab\,\mathbf{e}_y = -\frac{1}{c^2}\mathbf{m}\times\mathbf{v}\,.\end{aligned} \tag{11.23}$$

With point A as the origin there are contributions to the loop's magnetic dipole moment only from the parts BC and CD. On part BC the cross section area of the wire is modified by the length contraction, $\Delta' = \Delta/\gamma$, which implies that the current is reduced $I \to I/\gamma$. On part CD, however, there is not this effect, but the current is instead enhanced by the same amount, $I \to I\gamma$, as follows from the Lorentz transformation of the current density. This gives, for the magnetic dipole moment,

$$\begin{aligned}\mathbf{m}' &= \frac{1}{2}\frac{I}{\gamma}\int_0^b dy(a'\,\mathbf{e}_x + y\,\mathbf{e}_y)\times\mathbf{e}_y + \frac{1}{2}I\gamma\int_{a'}^0 dx(x\,\mathbf{e}_x + b\,\mathbf{e}_y)\times\mathbf{e}_x\\ &= \frac{1}{2}Iab(1+\frac{1}{\gamma^2})\mathbf{e}_z = \frac{1}{2}I\mathbf{a}\times\mathbf{b}\,(1-\frac{\beta^2}{2}) = (1-\frac{\beta^2}{2})\mathbf{m}\,.\end{aligned} \tag{11.24}$$

e) The arguments are as given above, with the current being $I\gamma$ in AB and CD and $I/\gamma$ in BC and DA.

f) To show that we have charge conservation, we focus on one of the corners (B), where the current running into the corner is different from the current running out of the corner. We consider the charge balance for an imagined box, which is at rest relative to reference frame $S'$, and which contains the corner B. For infinitesimal step $dt'$ in time, the difference in $S'$ between charge in and out of the box is

$$dQ' = (I'_{AB} - I'_{BC})dt = I(\gamma - \frac{1}{\gamma})dt = \gamma\beta^2 I dt\,. \tag{11.25}$$

This change of the charge inside the box can be explained by the fact that the corner B moves with velocity $v$ inside the box. This implies that the part of section AB which is inside the box increases with length $dx = vdt$ in the time interval $dt$. Since the charge of AB is $aIv/c^2$, the charge which enters the box in this time interval is

$$dQ' = \frac{aIv}{c^2}\frac{dx}{a'} = \gamma I \frac{v^2}{c^2} dt = \gamma\beta^2 I dt\,. \tag{11.26}$$

This agrees with the change of charge which follows from the total current that enters the box. Thus, the results are consistent with charge conservation.

### *Problem 11.4*

In this exercise the task is to fill in some details in the derivations of the force and torque on a stationary current in a magnetic field, discussed in the text book, Sect.11.3.

a) The first point is to prove the following identity, valid for stationary currents,

$$\int d^3r x_k j_l(\mathbf{r})(\mathbf{r}) x_l = \epsilon_{nkl} m_n\,, \tag{11.27}$$

where $m_n$ refers to components of the magnetic moment $\mathbf{m}$ of the current distribution.

From Eq. (11.43) in the textbook we derive the following identity

$$\int d^3r (x_k j_l(\mathbf{r}) + x_l j_k(\mathbf{r})) = 0 \quad \Rightarrow$$

$$\int d^3r x_k j_l(\mathbf{r}) = \frac{1}{2}\int d^3r (x_k j_l(\mathbf{r}) - x_l j_k(\mathbf{r}))\,. \tag{11.28}$$

In component form the magnetic dipole moment is defined as

$$m_s = \frac{1}{2}\epsilon_{skl}\int d^3r x_k j_l(\mathbf{r})\,, \tag{11.29}$$

which implies

$$\begin{aligned} \epsilon_{nkl} m_n &= \frac{1}{2}\epsilon_{nkl}\epsilon_{nrs}\int d^3r x_r j_s(\mathbf{r}) \\ &= \frac{1}{2}(\delta_{kr}\delta_{ls} - \delta_{ks}\delta_{lr})\int d^3r x_r j_s(\mathbf{r}) \\ &= \frac{1}{2}\int d^3r (x_k j_l(\mathbf{r}) - x_l j_k(\mathbf{r})) \\ &= \int d^3r x_k j_l(\mathbf{r}). \end{aligned} \tag{11.30}$$

b) We next start from the following expression, taken from the text book, for the leading term in the multipole expansion of the magnetic force,

$$(F_m)_i = \epsilon_{iln} \left( \int d^3 r x_k j_l(\mathbf{r}) \right) \partial_k B_n(0) \tag{11.31}$$

and show that it can be rewritten, in vector form, as

$$\mathbf{F}_m = \boldsymbol{\nabla}(\mathbf{m} \cdot \mathbf{B}). \tag{11.32}$$

We rewrite (11.31) by use of the result from a),

$$\begin{aligned} (F_m)_i &= \epsilon_{iln} \left( \int d^3 r\, x_k j_l(\mathbf{r}) \right) \partial_k B_n \\ &= \epsilon_{iln} \epsilon_{kls}\, m_s\, \partial_k B_n \\ &= (\delta_{ik}\delta_{ns} - \delta_{is}\delta_{nk})\, m_s\, \partial_k B_n \\ &= m_n \partial_i B_n - m_i \partial_n B_n \\ &= m_n \partial_i B_n\,, \end{aligned} \tag{11.33}$$

where we in the last step have made use of $\boldsymbol{\nabla} \cdot \mathbf{B} = 0$. In vector form this result reproduces Eq. (11.32).

c) The condition $\boldsymbol{\nabla} \times \mathbf{B} = 0$ means that $\partial_k B_l = \partial_l B_k$. We use this to rewrite the expression above as

$$(F_m)_i = m_n \partial_i B_n = m_n \partial_n B_i \tag{11.34}$$

which in vector form is

$$\mathbf{F}_m = (\mathbf{m} \cdot \boldsymbol{\nabla})\mathbf{B}\,. \tag{11.35}$$

d) We rewrite the expression for the torque (see Eq. (11.58) in the text book), by use of the identities (11.28) and (11.30),

$$\begin{aligned} (\tau_m)_i &= \epsilon_{iks}\epsilon_{sln} \left( \int d^3 r\, x_k j_l(\mathbf{r}) \right) B_n + ... \\ &= (\delta_{il}\delta_{kn} - \delta_{in}\delta_{kl}) \left( \int d^3 r\, x_k j_l(\mathbf{r}) \right) B_n + ... \\ &= \left( \int d^3 r\, x_n j_i(\mathbf{r}) \right) B_n - \left( \int d^3 r x_k j_k(\mathbf{r}) \right) B_i + ... \\ &= \left( \int d^3 r\, x_n j_i(\mathbf{r}) \right) B_n + ... \\ &= \epsilon_{ijn} m_j B_n + .... \end{aligned} \tag{11.36}$$

In vector form this result is

$$\boldsymbol{\tau}_m = \mathbf{m} \times \mathbf{B}. \tag{11.37}$$

## Chapter 12

# Electromagnetic radiation

***Problem 12.1***

An alternating current is running in an antenna of length $a$, oriented along the $z$-axis,

$$I(z,t) = I_0 \cos\left(\frac{\pi z}{a}\right) \cos \omega t\,, \quad -a/2 < z < a/2\,. \tag{12.1}$$

At time $t = 0$ the antenna is charge neutral, so that the linear charge density along the antenna vanishes, $\lambda(z, 0) = 0$.

a) Charge conservation is locally expressed in the form of the continuity equation

$$\frac{\partial \rho}{\partial t} + \boldsymbol{\nabla} \cdot \mathbf{j} = 0\,, \tag{12.2}$$

with $\rho$ as the volume density of the charge and $\mathbf{j}$ as the current density. With the current running in the $z$-direction we have

$$\boldsymbol{\nabla} \cdot \mathbf{j} = \frac{dj}{dz}\,, \quad j = j_z\,, \tag{12.3}$$

and integrated over the cross section of the antenna, we have the following expressions for the current and the linear charge density,

$$I = \int j dA\,, \quad \lambda = \int \rho dA\,. \tag{12.4}$$

This gives

$$\frac{\partial \lambda}{\partial t} + \frac{\partial I}{\partial z} = \int (\frac{\partial \rho}{\partial t} + \boldsymbol{\nabla} \cdot \mathbf{j}) dA = 0\,. \tag{12.5}$$

With the given form of the current we get

$$\frac{\partial \lambda}{\partial t} = -\frac{\partial I}{\partial z} = \frac{\pi I_0}{a} \sin\left(\frac{\pi z}{a}\right) \cos \omega t, \tag{12.6}$$

which, with the initial condition $\lambda(z, 0) = 0$, integrates to

$$\lambda(z,t) = \frac{\pi I_0}{\omega a} \sin\left(\frac{\pi z}{a}\right) \sin \omega t\,. \tag{12.7}$$

b) The electric dipole moment is

$$\mathbf{p} = \int \mathbf{r}\,\rho\, d^3r = \int_{-a/2}^{a/2} z\,\lambda(z,t)\,dz\,\mathbf{k}$$
$$= \frac{\pi I_0}{\omega a}\sin\omega t \int_{-a/2}^{a/2} z\sin\left(\frac{\pi z}{a}\right) dz\,\mathbf{k}\,. \tag{12.8}$$

Integration by parts gives

$$\int_{-a/2}^{a/2} z\sin\left(\frac{\pi z}{a}\right) dz = \left[-\frac{a}{\pi} z\cos\left(\frac{\pi z}{a}\right)\right]_{-a/2}^{a/2} + \frac{a}{\pi}\int_{-a/2}^{a/2} \cos\left(\frac{\pi z}{a}\right) dz$$
$$= (\frac{a}{\pi})^2\left[\sin\left(\frac{\pi z}{a}\right)\right]_{-a/2}^{a/2} = 2(\frac{a}{\pi})^2\,. \tag{12.9}$$

This gives for the electric dipole moment

$$\mathbf{p}(t) = 2\frac{aI_0}{\omega\pi}\sin\omega t\,\mathbf{k} \quad\Rightarrow\quad p_0 = 2\frac{aI_0}{\omega\pi}\,. \tag{12.10}$$

c) The expressions for electric dipole radiation fields are

$$\mathbf{B}(\mathbf{r},t) = -\frac{\mu_0}{4\pi c r}\mathbf{n}\times\ddot{\mathbf{p}}_{ret}\,, \quad \mathbf{n} = \mathbf{r}/r,$$
$$\mathbf{E}(\mathbf{r},t) = c\mathbf{B}(\mathbf{r},t)\times\mathbf{n} \tag{12.11}$$

with $\ddot{\mathbf{p}}_{ret}$ referring to $\ddot{\mathbf{p}}$ at the time $t_r = t - r/c$. In the present case we have

$$\ddot{\mathbf{p}}(t-\frac{r}{c}) = -\omega^2 p_0\sin(\omega t_r)\mathbf{k}\,. \tag{12.12}$$

For the fields on the $x$-axis, with $\mathbf{r} = r\mathbf{i}$, the magnetic field is

$$\mathbf{B}(r\mathbf{i},t) = -\frac{\mu_0\omega^2 p_0}{4\pi c r}\sin(\omega t_r)\,\mathbf{i}\times\mathbf{k}$$
$$= -\frac{\mu_0\omega a I_0}{2\pi^2 c r}\sin(\omega t_r)\,\mathbf{j}\,, \tag{12.13}$$

and the electric field is

$$\mathbf{E}(r\mathbf{i},t) = \frac{\mu_0\omega a I_0}{2\pi^2 r}\sin(\omega t_r)\,\mathbf{k}\,. \tag{12.14}$$

This shows that the radiation in the plane orthogonal to the antenna is linearly polarized (plane polarized), with the electric field oscillating along the direction of the antenna and the magnetic field oscillating in the direction orthogonal to the antenna and to the direction of propagation.

## *Problem 12.2*

We study a simple, classical model of the hydrogen atom, where the negatively charged electron moves in a circular orbit around the positively charged proton. Due to the large difference in mass of the two, the proton can be considered as having a fixed position. We regard the radius of the electron's orbit to be equal to the Bohr radius $a_0 = 0.53 \cdot 10^{-10}\,\mathrm{m}$. The electron has mass $m_e = 9.1 \cdot 10^{-31}\,\mathrm{kg}$ and a charge $e = -1.60 \cdot 10^{-19}\,\mathrm{C}$.We take the orbit plane to be the $x, y$-plane. The electric and magnetic constants are $\epsilon_0 = 8.85 \cdot 10^{-12}\,\mathrm{C^2N^{-1}m^{-2}}$ and $\mu_0 = 4\pi \cdot 10^{-7}\,\mathrm{N/A^2}$.

a) The equation of motion of the electron in the Coulomb field of the atomic nucleus is

$$m\mathbf{a} = e\mathbf{E} = -\frac{e}{4\pi\epsilon_0 r^3}\mathbf{r}\,. \tag{12.15}$$

For circular motion we have $\mathbf{a} = -\omega^2\mathbf{r}$, with $\omega$ as the angular velocity. Inserted in the equation of motion, this gives

$$m\omega^2 = \frac{e^2}{4\pi\epsilon_0 r^3}\,, \tag{12.16}$$

and with $r = a_0$ ($a_0$ here meaning the Bohr radius), this gives for the angular velocity,

$$\omega = \frac{e}{\sqrt{4\pi\epsilon_0 m a_0^3}} = 4.1 \cdot 10^{16}\,\mathrm{s}^{-1}\,. \tag{12.17}$$

b) Larmor's formula determines the radiation power,

$$P = \frac{\mu_0 e^2}{6\pi c}a^2 = \frac{\mu_0 e^2}{6\pi c}\omega^4 a_0^2 = 4.6 \cdot 10^{-8}\,\mathrm{W}\,. \tag{12.18}$$

c) The energy of the electron, expressed as a function of the radius of the circular orbit, is

$$\begin{aligned}\mathcal{E}(r) &= T + V = \frac{1}{2}mv^2 - \frac{e^2}{4\pi\epsilon_0 r}\\ &= \frac{1}{2}mr^2\omega^2 - \frac{e^2}{4\pi\epsilon_0 r} = -\frac{e^2}{8\pi\epsilon_0 r}\,.\end{aligned} \tag{12.19}$$

Assuming a slowly changing radius $r$, the time derivative of the energy is

$$\frac{d\mathcal{E}}{dt} = \frac{d\mathcal{E}}{dr}\frac{dr}{dt} = \frac{e^2}{8\pi\epsilon_0 r^2}\dot{r}\,. \tag{12.20}$$

With the effect of electromagnetic radiation taken into account, energy conservation implies

$$P + \frac{d\mathcal{E}}{dt} = 0 \quad\Rightarrow\quad \dot{r} = -\frac{8\pi\epsilon_0 r^2}{e^2}P\,. \tag{12.21}$$

Assuming $r = a_0$, this gives

$$\dot{r} = -\frac{4}{3}\frac{\omega^4 a_0^4}{c^3}, \tag{12.22}$$

and a rough estimate of the life time of this classical atom is then

$$T \approx \frac{a_0}{|\dot{r}|} = \frac{3}{4}\frac{c^3}{\omega^4 a_0^3} = 4.7 \cdot 10^{-11}\,\mathrm{s}. \tag{12.23}$$

***Problem 12.3***

A thin rigid rod of length $l$ rotates with constant angular velocity $\omega$ in a horizontal plane (the $x, y$-plane). At the two end points there are fixed charges with opposite signs, $+q$ and $-q$. This gives rise to a time dependent electric dipole moment

$$\mathbf{p}(t) = ql(\cos\omega t\,\mathbf{i} + \sin\omega t\,\mathbf{j}). \tag{12.24}$$

a) The magnetic field, in the case of electric dipole radiation, is

$$\mathbf{B}(\mathbf{r}, t) = \frac{\mu_0}{4\pi c r}\ddot{\mathbf{p}}_{ret} \times \mathbf{n}, \quad \mathbf{n} = \frac{\mathbf{r}}{r} \tag{12.25}$$

where $\ddot{\mathbf{p}}_{ret}$ is evaluated at the retarded time $t_r = t - \frac{r}{c}$. For the rotating electric dipole moment we have

$$\ddot{\mathbf{p}} = -\omega^2\mathbf{p} = -\omega^2 lq(\cos\omega t\,\mathbf{i} + \sin\omega t\,\mathbf{j}), \tag{12.26}$$

and the unit vector $\mathbf{n}$, when expressed in polar coordinates, is

$$\mathbf{n} = \sin\theta\cos\phi\,\mathbf{i} + \sin\theta\sin\phi\,\mathbf{j} + \cos\theta\,\mathbf{k}. \tag{12.27}$$

With these introduced in the expression for the magnetic field, we find

$$\begin{aligned}\mathbf{B}(\mathbf{r}, t) = &-\frac{\mu_0\omega^2 lq}{4\pi c r}(\cos\theta\sin\omega t_r\,\mathbf{i} \\ &- \cos\theta\cos\omega t_r\,\mathbf{j} - \sin\theta\sin(\omega t_r - \phi))\,\mathbf{k},\end{aligned} \tag{12.28}$$

which gives as amplitude of the oscillating field

$$B_0 = -\frac{\mu_0\omega^2 lq}{4\pi c r}. \tag{12.29}$$

The electric component of the radiation field is determined by the magnetic component as $\mathbf{E}(\mathbf{r}, t) = c\mathbf{B}(\mathbf{r}, t) \times \mathbf{n}$.

b) For radiation in the $x$-direction we have $\mathbf{n} = \mathbf{i}$, corresponding to polar angles $(\theta, \phi) = (\frac{\pi}{2}, 0)$. This gives

$$\mathbf{B}(\mathbf{r}, t) = -B_0 \sin \omega t_r \, \mathbf{k}\,, \tag{12.30}$$

which shows that the radiation is linearly polarized, with the $\mathbf{B}$-field oscillating in the $z$-direction.

For radiation in the $z$-direction we have $\mathbf{n} = \mathbf{k}$, corresponding to $\theta = 0$. In this case we have

$$\mathbf{B}(\mathbf{r}, t) = B_0(\sin \omega t_r \, \mathbf{i} - \cos \omega t_r \, \mathbf{j})\,, \tag{12.31}$$

which shows that the radiation is circularly polarized, with the $\mathbf{B}$-field (and the $\mathbf{E}$-field) rotating in the $x, y$-plane.

c) The energy density of the radiation is

$$\begin{aligned} u &= \frac{1}{2}(\epsilon_0 \mathbf{E}^2 + \frac{1}{\mu_0}\mathbf{B}^2) \\ &= \frac{1}{\mu_0} B_0^2 (\cos^2 \theta + \sin^2 \theta \sin^2(\omega t_r - \phi))\,. \end{aligned} \tag{12.32}$$

The time average $\overline{\sin^2 \omega t} = 1/2$ gives, for the averaged energy density,

$$\bar{u} = \frac{1}{2\mu_0} B_0^2 (1 + \cos^2 \theta)\,. \tag{12.33}$$

The energy density current is given by the Poynting vector

$$\mathbf{S} = \frac{1}{\mu_0} \mathbf{E} \times \mathbf{B} = \frac{c}{\mu_0} \mathbf{B}^2 \mathbf{n} = cu\mathbf{n}\,. \tag{12.34}$$

It has its maximal value in the direction where the energy density $\bar{u}$ is largest. This happens when $\cos\theta = \pm 1$, which means in the positive or negative direction along the $z$-axis.

### *Problem 12.4*

In a circular loop of radius $a$, an oscillating current of the form $I = I_0 \cos \omega t$ is running. The current loop lies in the $x, y$-plane, with the center of the loop at the origin. The loop is at all times charge neutral, and the following inequality, $a\,\omega << c$, is satisfied.

a) We evaluate the magnetic dipole moment of the current, defined as

$$\mathbf{m} = \frac{1}{2} \int (\mathbf{r} \times \mathbf{j}) dV\,, \tag{12.35}$$

where $\mathbf{j}$ is the current density (inside the conducting wire). Due to the circular motion we write it as $\mathbf{j} = j\mathbf{e}_\phi$, with $\mathbf{e}_\phi$ as the unit vector tangential to the current loop. With the current $I$ as the integral of $j$ over the cross section of the current wire, and $\mathbf{e}_r$ as the unit vector in the radial direction, the magnetic moment is

$$\begin{aligned}\mathbf{m} &= \frac{1}{2}a^2 I \int_0^{2\pi} (\mathbf{e}_r \times \mathbf{e}_\phi) d\phi \\ &= \pi a^2 I \mathbf{e}_z \\ &= \pi a^2 I_0 \cos\omega t \, \mathbf{e}_z \,. \end{aligned} \tag{12.36}$$

This shows that the amplitude of the oscillations of $\mathbf{m}$ is $m_0 = \pi a^2 I_0$.

When $a\,\omega << c$, which corresponds to $\lambda >> a$ where $\lambda$ is the (typical) wave length of the radiation, the radiation field will be dominated by the lowest multipole terms. In the present case, since the current loop is charge neutral, both the electric dipole and quadrupole terms vanish, and the magnetic dipole term is therefore the dominant one.

b) The expression given above for $\mathbf{m}$ gives, for $\ddot{\mathbf{m}}$,

$$\ddot{\mathbf{m}}_{ret} = -\pi a^2 \omega^2 I_0 \cos(\omega(t - \frac{r}{c}))\, \mathbf{e}_z \,. \tag{12.37}$$

With $\omega(t - \frac{r}{c}) = \omega t - kr$, where $k = \omega/c$, the electric and magnetic fields are then determined as

$$\begin{aligned}\mathbf{E}(\mathbf{r}, t) &= -\frac{\mu_0}{4\pi c r} \ddot{\mathbf{m}}_{ret} \times \mathbf{n} \\ &= \frac{\mu_0 a^2}{4cr} I_0 \omega^2 \cos(\omega t - kr) \sin\theta \, \mathbf{e}_\phi \,, \end{aligned} \tag{12.38}$$

and

$$\begin{aligned}\mathbf{B}(\mathbf{r}, t) &= -\frac{1}{c}\mathbf{E}(\mathbf{r}, t) \times \mathbf{n} \\ &= -\frac{\mu_0 a^2}{4cr} I_0 \omega^2 \cos(\omega t - kr) \sin\theta \, \mathbf{e}_\theta \,. \end{aligned} \tag{12.39}$$

Here we have used the relations $\mathbf{e}_z \times \mathbf{n} = \sin\theta \, \mathbf{e}_\phi$, $\mathbf{e}_\phi \times \mathbf{n} = \mathbf{e}_\theta$. The expressions for $\mathbf{E}$ and $\mathbf{B}$ show that they define a monochromatic wave with angular frequency $\omega$ which propagates in the radial direction, away from the magnetic dipole. Since the $\mathbf{E}$ and $\mathbf{B}$ fields oscillate in fixed directions, along the unit vectors $\mathbf{e}_\phi$ and $\mathbf{e}_\theta$, respectively, the radiation is linearly polarized.

c) The radiated power per unit solid angle is expressed in terms of the Poynting vector, $\mathbf{S} = \frac{1}{\mu_0}\mathbf{E} \times \mathbf{B}$, as

$$\frac{dP}{d\Omega} = r^2 S = \frac{1}{\mu_0 c}\mathbf{E}^2 = \frac{\mu_0 a^4}{16 c^3 r^2} I_0^2 \omega^4 \cos^2(\omega t - kr) \sin^2\theta \,. \tag{12.40}$$

Averaged over time gives $\cos^2 \omega t \to 1/2$, and integrated over angles, the above expression gives, for the radiated power,

$$\bar{P} = \int \frac{d\bar{P}}{d\Omega} d\Omega = 2\pi \frac{\mu_0 a^4}{32 c^3} I_0^2 \omega^4 \int \sin^3 \theta d\theta = \frac{\mu_0 \pi a^4}{12 c^3} I_0^2 \omega^4 . \tag{12.41}$$

d) The induced electric field in the secondary current loop is, according to Faraday's law of induction, determined by the time derivative of the magnetic flux through the loop. With the infinitesimal area vector of the loop written as $d\mathbf{A} = \mathbf{e}\, dA$, where $\mathbf{e}$ is a unit vector perpendicular to the second loop, the loop integral of the electric field is

$$\oint_C \mathbf{E} \cdot d\mathbf{s} = \frac{d}{dt} \int_S \mathbf{B} \cdot \mathbf{e}\, dA . \tag{12.42}$$

This shows that the induced field will be strongest when $|\mathbf{B} \cdot \mathbf{e}|$ is maximal. For points in the $x, y$-plane the $\mathbf{B}$ field is oriented along the $z$-axis. This implies that the signal received by the secondary loop is maximal for the orientations, $\mathbf{e} = \pm \mathbf{e}_z$.

### *Problem 12.5*

An electron with charge $e$ and mass $m$ is moving with constant speed in a circle under the influence of a constant magnetic field $\mathbf{B}_0$. The magnetic field is directed along the $z$-axis, while the motion of the electron takes place in the $x, y$ plane. Since the electron is accelerated, it will radiate electromagnetic energy and thereby lose kinetic energy. We study this effect, assuming that the speed of the electron is non-relativistic.

a) For the circular motion we have the following relations between the velocity $v$, the acceleration $a$ and the radius $r$ of the circle,

$$v = \omega r , \quad a = \omega v = \omega^2 r , \tag{12.43}$$

where $\omega = -eB_0/m$. With this applied to Larmor's radiation formula, we find that radiation power is

$$P = \frac{\mu_0 e^2}{6\pi c} a^2 = \frac{\mu_0 e^2}{6\pi c} \omega^4 r^2 . \tag{12.44}$$

b) Energy conservation applied to the electron gives

$$\frac{d}{dt}(\frac{1}{2} m v^2) = -P \quad \Rightarrow \quad \frac{1}{2} m \omega^2 \frac{d}{dt} r^2 = -\frac{\mu_0 e^2}{6\pi c} \omega^4 r^2 . \tag{12.45}$$

This can be written in the form

$$\frac{d}{dt} r^2 = -2\lambda r^2 , \quad \lambda = \frac{\mu_0 e^2}{6\pi c} \frac{\omega^2}{m} . \tag{12.46}$$

By rewriting the differential equation as

$$\frac{dr^2}{r^2} = -2\lambda dt\,, \tag{12.47}$$

the equation can easily be integrated to give the solution

$$\ln r^2 = -2\lambda t + \ln r_0^2\,, \tag{12.48}$$

with the last term as an integration constant. From this follows

$$r = r_0 e^{-\lambda t} = r_0 \exp\left(-\frac{\mu_0 e^2\omega^2}{6\pi mc}t\right). \tag{12.49}$$

***Problem 12.6***

An antenna is composed of two parts. One part is a linear antenna along the $z$-axis, with end points $z = \pm a/2$. It carries the current

$$I_1 = I_0 \sin\omega t \cos\left(\frac{\pi z}{a}\right). \tag{12.50}$$

The other part is a circular antenna, which lies in the $x, y$-plane, and is centered at the origin of the coordinate system. It has radius $2a$ and carries the current

$$I_2 = I_0 \sin\omega t\,. \tag{12.51}$$

a) The electric dipole moment is

$$\mathbf{p} = \int_{-a/2}^{a/2} z\lambda(z,t)dz\,\mathbf{k}\,, \tag{12.52}$$

with $\lambda(z,t)$ as the linear charge density along the $z$-axis. We exploit the continuity equation of charge in the form

$$\frac{\partial\lambda}{\partial t} + \frac{\partial I_1}{\partial z} = 0\,, \tag{12.53}$$

and find for the time derivative of the electric dipole moment,

$$\begin{aligned}\dot{\mathbf{p}} &= \int_{-a/2}^{a/2} z\frac{\partial\lambda}{\partial t}dz\,\mathbf{k} \\ &= -\int_{-a/2}^{a/2} z\frac{\partial I_1}{\partial z}dz\,\mathbf{k} \\ &= \int_{-a/2}^{a/2} I_1(z,t)dz - [zI_1]_{-a/2}^{a/2}\,\mathbf{k}\,.\end{aligned} \tag{12.54}$$

Since the current vanishes at the endpoints of the conducting wire there is no contribution from the last term, and integration of the first term gives

$$\begin{aligned}\dot{\mathbf{p}} &= \frac{a}{\pi} I_0 \sin\omega t \left[\sin\left(\frac{\pi z}{a}\right)\right]_{-a/2}^{a/2} \mathbf{k} \\ &= \frac{2a}{\pi} I_0 \sin\omega t\, \mathbf{k}\,. \end{aligned} \tag{12.55}$$

The magnetic dipole moment is given by

$$\begin{aligned}\mathbf{m} &= \frac{1}{2}\oint I_2 \mathbf{r} \times d\mathbf{r} \\ &= \frac{1}{2} I_0 \sin\omega t (2a)^2 \int_0^{2\pi} d\phi\, \mathbf{k} \\ &= 4\pi a^2 I_0 \sin\omega t\, \mathbf{k}\,. \end{aligned} \tag{12.56}$$

b) The dipole contributions to the electric and magnetic radiation fields are given by

$$\begin{aligned}\mathbf{E}(\mathbf{r}, t) &= \frac{\mu_0}{4\pi r}((\ddot{\mathbf{p}} \times \mathbf{n}) \times \mathbf{n} - \frac{1}{c}\ddot{\mathbf{m}} \times \mathbf{n} + ...)_{ret}\,, \\ \mathbf{B}(\mathbf{r}, t) &= -\frac{1}{c}\mathbf{E}(\mathbf{r}, t) \times \mathbf{n}\,, \end{aligned} \tag{12.57}$$

with $\mathbf{n} = \mathbf{r}/r$. The double time derivatives of $\mathbf{p}$ and $\mathbf{m}$, which follow from the expressions above, are

$$\ddot{\mathbf{p}} = \frac{2a}{\pi}\omega I_0 \cos\omega t\, \mathbf{k}\,, \qquad \ddot{\mathbf{m}} = -4\pi a^2 \omega^2 I_0 \sin\omega t\, \mathbf{k}\,. \tag{12.58}$$

We introduce the orthogonal unit vectors $\mathbf{e}_1$ and $\mathbf{e}_2$ through the equations

$$\mathbf{k} \times \mathbf{n} = \sin\theta\, \mathbf{e}_1\,, \qquad (\mathbf{k} \times \mathbf{n}) \times \mathbf{n} = \sin\theta\, \mathbf{e}_2\,, \tag{12.59}$$

which means that $(\mathbf{n}, \mathbf{e}_1, \mathbf{e}_2)$ defines an orthonormalized basis set in three-dimensional space.

By use of the expressions for the $\mathbf{E}$ and $\mathbf{B}$ fields, we find the following expression for Poynting's vector,

$$\begin{aligned}\mathbf{S} &= \frac{1}{\mu_0}\mathbf{E} \times \mathbf{B} = \frac{1}{\mu_0 c}\mathbf{E}^2 \mathbf{n} \\ &= \frac{\mu_0}{16\pi^2 r^2 c}(\ddot{\mathbf{p}}^2 + \frac{1}{c^2}\ddot{\mathbf{m}}^2) \sin^2\theta\, \mathbf{n}\,. \end{aligned} \tag{12.60}$$

The time averages of the contributions to the radiation power from the electric and magnetic dipole moments are assumed to be equal. That gives

$$\overline{\ddot{\mathbf{p}}^2} = \frac{1}{c^2}\overline{\ddot{\mathbf{m}}^2} \quad \Rightarrow \quad \omega = \frac{c}{2\pi^2 a}\,. \tag{12.61}$$

Poynting's vector in this case is time independent and given by

$$\mathbf{S} = \frac{\mu_0}{16\pi^2 r^2 c}\frac{4a^2}{\pi^2}\left(\frac{c}{2\pi^2 a}\right)^2 I_0^2 \sin^2\theta\,\mathbf{n}$$
$$= \frac{\mu_0 c}{16\pi^8 r^2} I_0^2 \sin^2\theta\mathbf{n}\,, \tag{12.62}$$

and the radiation power per unit solid angle is

$$\frac{dP}{d\Omega} = r^2 S = \frac{\mu_0 c}{16\pi^8} I_0^2 \sin^2\theta\,. \tag{12.63}$$

c) With equal contributions from the electric and magnetic dipole terms, the electric field is

$$\mathbf{E}(\mathbf{r},t) = \frac{\mu_0 c}{4\pi^4 r} I_0 \sin\theta(\sin\omega t\,\mathbf{e}_1 + \cos\omega t\,\mathbf{e}_2)\,. \tag{12.64}$$

The electric vector rotates in the plane spanned by $\mathbf{e}_1$ and $\mathbf{e}_2$. Since the rotation is clockwise around the direction of propagation $\mathbf{n}$, the polarization is left-handed, circular. If the contributions from the electric and magnetic dipole terms are not equal, the coefficients before $\sin\omega t$ and $\cos\omega t$ in the expressions for $\mathbf{E}$ will no longer be equal. The radiation will then be elliptically polarized, with $\mathbf{e}_1$ and $\mathbf{e}_2$ defining the symmetry axes of the ellipse.

### *Problem 12.7*

An electron, with charge $e$ and mass $m$, performs oscillations in an electromagnetic field with the following components,

$$E_x = E_0\cos(kz - \omega t)\,, \quad E_y = E_z = 0\,. \tag{12.65}$$

a) For a monochromatic plane wave the magnetic field is related to the electric field by

$$\mathbf{B} = \frac{1}{c}\mathbf{n}\times\mathbf{E}\,, \quad \mathbf{n} = \mathbf{k}/k\,. \tag{12.66}$$

In the present case, with the electromagnetic wave propagating in the $z$-direction, this gives

$$B_y = \frac{1}{c}E_x = \frac{E_0}{c}\cos(kx - \omega t)\,, \quad B_x = B_z = 0\,. \tag{12.67}$$

The Poynting vector is then

$$\mathbf{S} = \frac{1}{\mu_0}\mathbf{E}\times\mathbf{B}$$
$$= \epsilon_0 c\,E_x^2\,\mathbf{e}_z$$
$$= \epsilon_0 c\,E_0^2\cos^2(kz - \omega t)\,\mathbf{e}_z\,. \tag{12.68}$$

b) The assumed motion of the electron is

$$\dot{x} = -\frac{eE_0}{m\omega}\sin(kz - \omega t)\,, \quad \dot{y} = \dot{z} = 0\,. \tag{12.69}$$

This gives, for the acceleration,

$$\ddot{x} = \frac{e\mathbf{E}_0}{m}\cos(kz - \omega t)\,, \;\; \ddot{y} = \ddot{z} = 0 \quad \Rightarrow \quad m\mathbf{a} = e\mathbf{E}\,. \tag{12.70}$$

This equation is seen to be correct if we can neglect the magnetic force $e\mathbf{v} \times \mathbf{B}$ and assume non-relativistic motion. The magnetic force is negligible relative to the electric force if $vB << E$, and since for the plane wave we have $B = E/c$, this is satisfied if the motion is non-relativistic, $v << c$. With the motion of the electron given by (12.69), this can further be expressed as a constraint on the ratio between the amplitude and the frequency of the electromagnetic wave,

$$\frac{eE_0}{mc\,\omega} << 1\,. \tag{12.71}$$

c) The time averaged emitted power from the electron is determined by Larmor's formula,

$$\begin{aligned} \bar{P}_{rad} &= \frac{\mu_0 e^2}{6\pi c}\overline{\mathbf{a}^2} \\ &= \frac{\mu_0 e^4}{6\pi m^2 c}\overline{\mathbf{E}^2} \\ &= \frac{\mu_0 e^4}{12\pi m^2 c}E_0^2\,. \end{aligned} \tag{12.72}$$

The time averaged energy current density of the plane wave is

$$\bar{S}_{pw} = \frac{1}{\mu_0 c}\overline{\mathbf{E}^2} = \frac{1}{2\mu_0 c}E_0^2 \tag{12.73}$$

with $\mathbf{S}_{pw}$ as the Poynting vector of the plane wave. This gives for the interaction cross section

$$\sigma = \frac{\bar{P}_{rad}}{\bar{S}_{pw}} = \frac{\mu_0^2 e^4}{6\pi m^2}\,. \tag{12.74}$$

d) We assume in the following that the electron oscillates about the origin, $\mathbf{r} = 0$. We denote the Poynting vector of the radiation field from the electron by $\mathbf{S}_{rad}$. The corresponding radiated power per unit solid angle is

$$\frac{dP_{rad}}{d\Omega} = r^2\,\mathbf{S}_{rad} \cdot \mathbf{n}\,, \quad \mathbf{n} = \frac{\mathbf{r}}{r}\,. \tag{12.75}$$

We make use of the following expression for the magnetic component of the radiation field far from the electron

$$\mathbf{B}_{rad}(\mathbf{r},t) = \frac{\mu_0 e}{4\pi c}\left[\frac{\mathbf{a}\times\mathbf{n}}{r}\right]_{ret}. \tag{12.76}$$

The Poynting vector is then

$$\begin{aligned}\mathbf{S}_{rad} &= \frac{1}{\mu_0}\mathbf{E}_{rad}\times\mathbf{B}_{rad} = \frac{c}{\mu_0}\mathbf{B}_{rad}^2\mathbf{n} \\ &= \frac{c}{\mu_0}\frac{\mu_0^2 e^2}{16\pi^2 c^2}\left[\frac{\mathbf{a}\times\mathbf{n}}{r}\right]_{ret}^2\mathbf{n} \\ &= \frac{\mu_0 e^2}{16\pi^2 c r^2}(\mathbf{a}^2 - (\mathbf{a}\cdot\mathbf{n})^2)_{ret}\,\mathbf{n}\,. \end{aligned} \tag{12.77}$$

We introduce $\theta$ as the angle between the $x$-axis and $\mathbf{n}$. This gives

$$\begin{aligned}(\mathbf{a}^2 - (\mathbf{a}\cdot\mathbf{n})^2)_{ret} &= a_{ret}^2(1-\cos^2\theta) \\ &= \frac{e^2}{m^2}E_0^2\sin^2\theta\cos^2(kz-\omega t_{ret})\,. \end{aligned} \tag{12.78}$$

The time averaged differential power is then

$$\frac{d\bar{P}}{d\Omega} = \overline{r^2\mathbf{S}_{rad}\cdot\mathbf{n}} = \frac{\mu_0 e^4}{32\pi^2 m^2 c}E_0^2\sin^2\theta\,. \tag{12.79}$$

The radiation is maximal when $\theta = \pi/2$, that is in directions perpendicular to the $x$-axis, which is the direction of oscillations of the electron. The radiation is minimal for $\theta = 0$ or $\theta = \pi$, that is in the directions of the $x$-axis.

The electric component of the radiation field is

$$\begin{aligned}\mathbf{E}_{rad}(\mathbf{r},t) &= \frac{\mu_0 e}{4\pi r}\left[(\mathbf{a}\times\mathbf{n})\times\mathbf{n}\right]_{ret} \\ &= \frac{\mu_0 e^2}{4\pi m r}E_0\cos(kz-\omega t_{ret})(\mathbf{e}_x\times\mathbf{n})\times\mathbf{n}\,, \end{aligned} \tag{12.80}$$

which shows that the radiation is linearly polarized in the direction of $(\mathbf{e}_x\times\mathbf{n})\times\mathbf{n}$.

### *Problem 12.8*

A linear antenna of length $2a$ is oriented along the $z$-axis, with its center at the origin. The assumption is that the charge of the antenna is at all times located at the endpoints, and the current in the antenna is given by $I = I_0 \sin \omega t$. The antenna is electrically neutral at time $t = 0$. The radiation from the antenna can be treated as electric dipole radiation. The spherical coordinates of point $A$ are $(r, \theta, \phi)$, and the corresponding orthonormal vector basis is $\{\mathbf{e}_r, \mathbf{e}_\theta, \mathbf{e}_\phi\}$.

a) We refer to the endpoint charges as $Q(\pm a, t) = \pm Q(t)$, with $Q(0) = 0$. Charge conservation implies

$$\begin{aligned}
\frac{dQ}{dt} &= I = I_0 \sin \omega t \\
\Rightarrow \quad Q(t) &= \int_0^t I_0 \sin \omega t \, dt \\
&= \left[ -\frac{I_0}{\omega} \cos \omega t \right]_0^t \\
&= \frac{I_0}{\omega}(1 - \cos \omega t)\,. \qquad (12.81)
\end{aligned}$$

The electric dipole moment is then

$$\begin{aligned}
&\mathbf{p}(t) = 2Q(t)a\,\mathbf{k} = \frac{2aI_0}{\omega}(1 - \cos \omega t)\mathbf{k} \\
\Rightarrow \quad &\ddot{\mathbf{p}} = 2I_0 \omega a \cos \omega t\, \mathbf{k}\,. \qquad (12.82)
\end{aligned}$$

b) For electric dipole radiation, the magnetic component of the field is

$$\begin{aligned}
\mathbf{B}(\mathbf{r}, t) &= \frac{\mu_0}{4\pi r c} \ddot{\mathbf{p}}_{ret} \times \mathbf{n} \\
&= \frac{\mu_0 I_0 \omega a}{2\pi r c} \cos \omega t_r \, \mathbf{k} \times \mathbf{e}_r \\
&= \frac{\mu_0 I_0 \omega a}{2\pi r c} \sin\theta \cos \omega t_r \, \mathbf{e}_\phi\,, \qquad (12.83)
\end{aligned}$$

with $t_r = t - r/c$. The electric component of the field is

$$\mathbf{E}(\mathbf{r}, t) = c\mathbf{B}(\mathbf{r}, t) \times \mathbf{e}_r = \frac{\mu_0 I_0 \omega a}{2\pi r} \sin\theta \cos \omega t_r \, \mathbf{e}_\theta\,. \qquad (12.84)$$

The radiation is linearly polarized, since $\mathbf{E}$ oscillates with time in the fixed direction defined by $\mathbf{e}_\theta$. Similarly $\mathbf{B}$ oscillates in the orthogonal direction, given by $\mathbf{e}_\phi$.

c) We write the fields as $\mathbf{B} = B\mathbf{e}_\phi$ and $\mathbf{E} = cB\mathbf{e}_\theta$. The Poynting vector is

$$\begin{aligned}\mathbf{S} &= \frac{1}{\mu_0}\mathbf{E}\times\mathbf{B}\\ &= \frac{c}{\mu_0}B^2\mathbf{e}_\theta\times\mathbf{e}_\phi = \frac{c}{\mu_0}B^2\mathbf{e}_r\\ &= \frac{\mu_0\omega^2a^2}{4\pi^2r^2c}I_0^2\sin^2\theta\cos^2\omega t_r\,\mathbf{e}_r\,. \end{aligned} \tag{12.85}$$

The radiation power per unit solid angle is then

$$\begin{aligned}\frac{dP}{d\Omega} &= r^2S = \frac{c}{\mu_0}r^2B^2\\ &= \frac{\mu_0\omega^2a^2}{4\pi^2c}I_0^2\sin^2\theta\cos^2\omega t_r\,, \end{aligned} \tag{12.86}$$

and averaged over time the result is

$$\overline{\frac{dP}{d\Omega}} = \frac{\mu_0\omega^2a^2}{8\pi^2c}I_0^2\sin^2\theta\,. \tag{12.87}$$

The integration over angles is

$$\int d\Omega\sin^2\theta = 2\pi\int_0^\pi\sin^3\theta d\theta = \frac{8}{3}\pi\,, \tag{12.88}$$

which for the averaged total power gives

$$\bar{P} = \frac{\mu_0\omega^2a^2}{3\pi c}I_0^2 \equiv \frac{1}{2}RI_0^2 \quad\Rightarrow\quad R = \frac{2\mu_0\omega^2a^2}{3\pi c}\,. \tag{12.89}$$

With $R_0$ as regular resistance and $R$ as radiation resistance, the total power consumed by the antenna is

$$\bar{P}_{tot} = \frac{1}{2}(R+R_0)I_0^2\,. \tag{12.90}$$

d) With specified values, $2a = 5\,\mathrm{cm}$, frequency $f = 150\,\mathrm{MHz}$ and current $I_0 = 30\,\mathrm{A}$, the radiation resistance and the time averaged radiation power are

$$R = \frac{8\pi\mu_0f^2a^2}{3c} = 0.49\,\Omega\,,\quad \bar{P} = \frac{1}{2}RI_0^2 = 222\,\mathrm{W}\,. \tag{12.91}$$

*Problem 12.9*

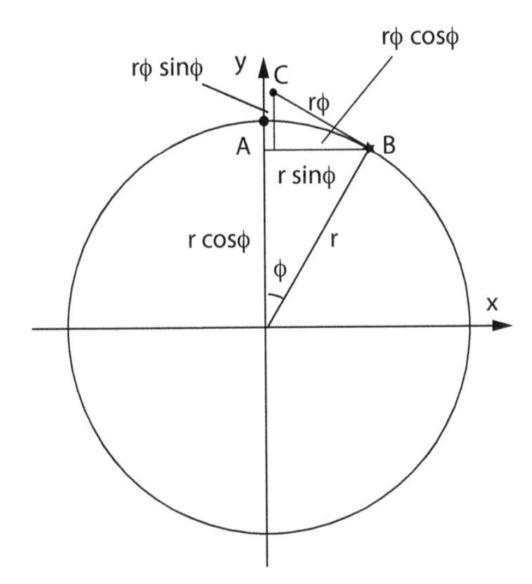


Fig. 12.1 Geometric construction of the coordinates of the radiation point $C$ in Problem 12.9.

A charged particle moves with ultrarelativistic speed in a circular orbit. The radiation which is caused by acceleration of the charge, will almost exclusively be sent in the forward direction, tangential to the circular orbit. At time $t = 0$ the particle is located at a point $A$. At an earlier time, $t_B < 0$, the radiation is emitted from the particle at a point $B$. The angle $\phi$ between $A$ and $B$ on the circle is given by $\phi = c|t_B|/r$, with $r$ as the radius of the circle, when we approximate the speed of the particle with the speed of light. The radiation radiated from point $B$ will at time $t = 0$ have reached a third point $C$.

a) The coordinates of $C$ can be found by a simple geometric construction. As shown in Fig. 12.1, point $C$ can be reached from the origin by use of two congruent, right-angled triangles. The upper one is scaled relative to the lower one by the factor $\phi$. The corresponding sides of the two are also different in orientation, corresponding to a reflection $x \leftrightarrow y$. The result for the coordinates is as stated in the text of the problem,

$$x = r(\sin\phi - \phi\cos\phi)\,, \quad y = r(\cos\phi + \phi\sin\phi). \tag{12.92}$$

b) Fig. 12.2b shows the spatial position of the radiation, close to the charged particle at time $t = 0$. The radiation in this case has been emitted in a small angular interval $0 < \phi \lesssim \pi/10$. The radiation forms the curved, almost vertical curve, and the direction of propagation is indicated by the arrow. A part of the circular orbit of the particle, which in the figure is located close to the $y$-axis, is shown by the slightly curved, almost horizontal curve.

c) Fig. 12.2c shows the location of the radiation that has been emitted by the particle in the larger angular interval $0 < \phi < 2\pi$. In this figure the full circle of the particle orbit is seen, and the spiral like form of the radiation from the particle is apparent. Here, the direction of propagation of the radiation is also shown.

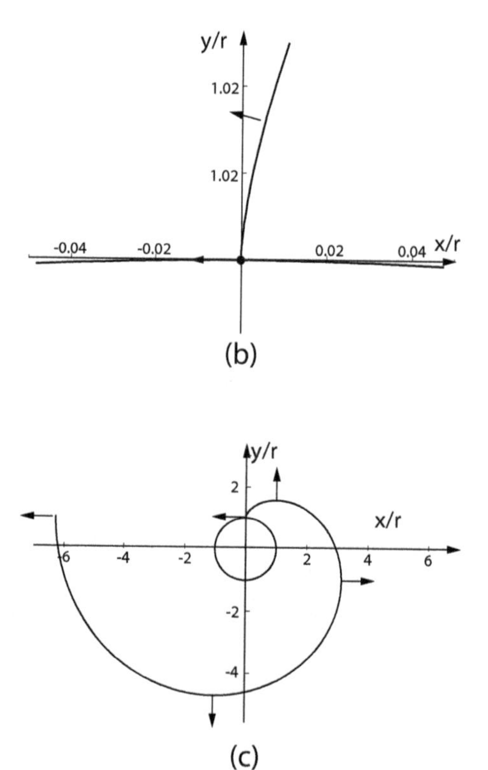


Fig. 12.2 The location of the synchrotron radiation at the instant when the charged particle passes the positive $y$-axis. (b) shows the radiation close to the particle, while (c) shows the radiation, at the same instant, which has been emitted from the charge during a full period of circulation.

# PART 4

# Classical Field Theory

# Chapter 13

# Lagrangian and Hamiltonian formulations

### *Problem 13.1*

This is an electrostatic situation, where $\rho$ is the electric charge distribution and $\phi$ is the electric potential. The Lagrangian density has the form,

$$\mathcal{L} = \frac{1}{2}(\nabla\phi)^2 - \frac{\rho}{\epsilon_0}\,\phi. \tag{13.1}$$

The problem is to determine the derivatives

$$\frac{\partial\mathcal{L}}{\partial\phi}, \quad \frac{\partial\mathcal{L}}{\partial\phi_{,i}} \tag{13.2}$$

to find the corresponding Euler–Lagrange equation and to check that the result is in accordance with Poisson's equation,

$$\nabla^2\phi = -\frac{\rho}{\epsilon_0}. \tag{13.3}$$

The Lagrange density is

$$\mathcal{L} = \frac{1}{2}\phi_{,i}\phi^{,i} - \frac{\rho}{\epsilon_0}\,\phi, \tag{13.4}$$

which gives

$$\frac{\partial\mathcal{L}}{\partial\phi} = -\frac{\rho}{\epsilon_0}\,\phi, \quad \frac{\partial\mathcal{L}}{\partial\phi_{,i}} = \phi^{,i} \tag{13.5}$$

and the Euler–Lagrange equation,

$$\frac{\partial\mathcal{L}}{\partial\phi} - \frac{\partial}{\partial x^{,i}}\left(\frac{\partial\mathcal{L}}{\partial\phi_{,i}}\right) = -\frac{\rho}{\epsilon_0}\,\phi - \frac{\partial}{\partial x^{,i}}\phi^{,i} = 0. \tag{13.6}$$

The result is

$$\frac{\partial^2}{\partial x^{,i}\partial x_{,i}}\phi = -\frac{\rho}{\epsilon_0}, \tag{13.7}$$

which we simplify to

$$\nabla^2\phi = -\frac{\rho}{\epsilon_0}. \tag{13.8}$$

This is in accordance with Poisson's equation.

### *Problem 13.2*

(a) The Lagrangian density with $\phi^4$-theory is

$$\mathcal{L} = \frac{1}{2}(\partial_t\phi)^2 - \frac{1}{2}(\partial_x\phi)^2 - \frac{1}{4}\lambda\left(\phi^2 - \rho_0\right)^2, \tag{13.9}$$

where $c = 1$, and $\lambda$ and $\rho_0$ are constants.

The Euler–Lagrange equation is found in the following way:

$$\frac{\partial\mathcal{L}}{\partial\phi} - \frac{\partial}{\partial t}\frac{\partial\mathcal{L}}{\partial\dot{\phi}} - \frac{\partial}{\partial x}\frac{\partial\mathcal{L}}{\partial\phi'} = 0$$

$$\Rightarrow -\ddot{\phi} + \phi'' - \lambda\phi(\phi^2 - \rho_0) = 0 \tag{13.10}$$

When $\phi(t,x)$ solves the equation, so does $-\phi(t,x)$, since the equation is symmetric under $\phi \to -\phi$.

(b) The energy density is

$$\begin{aligned} T^{00} &= -\mathcal{L} + \frac{\partial\mathcal{L}}{\partial\dot{\phi}}\dot{\phi} \\ &= -\mathcal{L} + \dot{\phi}^2 \\ &= \frac{1}{2}(\dot{\phi}^2 + \phi'^2) + \frac{1}{4}\lambda(\phi^2 - \rho_0)^2. \end{aligned} \tag{13.11}$$

(c) A classical vacuum state is

$$\phi^2 = \rho_0 \quad \Rightarrow \quad T^{00} = 0, \tag{13.12}$$

which is a minimum for $T^{00}$. There are in fact two possible values:

$$\phi_\pm = \pm\sqrt{\rho_0}. \tag{13.13}$$

(d) The problem now is to show that the following is a static solution to the Euler–Lagrange equation (with $\ddot{\phi} = 0$):

$$\begin{aligned} \phi_\pm(x) &= \pm\sqrt{\rho_0}\tanh\left[\sqrt{\frac{\lambda\rho_0}{2}}(x-a)\right] \qquad a = \text{const.} \\ &= \pm\sqrt{\rho_0}\tanh\, y, \quad y = \sqrt{\frac{\lambda\rho_o}{2}}(x-a). \end{aligned} \tag{13.14}$$

We study separately the two parts of the equation:

$$\begin{aligned} &(1) \quad \phi'' = \frac{d^2}{dx^2}\{\sqrt{\rho_0}\tanh\, y\}, \\ &(2) \quad \lambda\phi(\phi^2 - \rho_0) = \lambda\rho_0\sqrt{\lambda\rho_0}\tanh\, y\tanh^2 y. \end{aligned} \tag{13.15}$$

We make use of this:

$$(1)\ \frac{d}{dx}\tanh\ y = \frac{dy}{dx}\frac{d}{dy}\tanh\ y = \sqrt{\frac{\lambda\rho_0}{2}}\frac{1}{\cosh^2 y}$$

$$\Rightarrow\quad \frac{\partial^2}{\partial x^2}\tanh\ y = \sqrt{\frac{\lambda\rho_0}{2}}\frac{d}{dy}\frac{1}{\cosh^2 y} = -\lambda\rho_0\frac{\sinh^2 y}{\cosh^3 y}, \tag{13.16}$$

$$\begin{aligned}(2)\quad \lambda\phi(\phi^2-\rho_0) &= \lambda\rho_0\sqrt{\rho_0}\tanh\ y(\tanh^2 y - 1)\\ &= \lambda\rho_0\sqrt{\rho_0}\ \frac{\sinh\ y}{\cosh\ y}\left(\frac{\sinh^2 y}{\cosh^2 y} - 1\right)\\ &= \lambda\rho_0\sqrt{\rho_0}\ \frac{\sinh\ y}{\cosh^3 y}.\end{aligned} \tag{13.17}$$

This shows that (1) + (2) = 0, which implies that Eq. (13.14) is satisfied:

$$\pm\,\phi(x) = \pm\sqrt{\rho_0}\tanh\left[\sqrt{\frac{\lambda\rho_0}{2}}(x-a)\right], \tag{13.18}$$

and that Euler–Lagrange equation is therefore confirmed.

(e) To find the energy density for the static Euler–Lagrange equation, we start as follows:

$$T^{00} = \frac{1}{2}\phi'^2 + \frac{1}{4}\lambda(\phi^2-\rho_0)^2. \tag{13.19}$$

The derivative $\phi'$ is the next step:

$$\begin{aligned}\phi' &= \frac{\partial}{\partial x}\sqrt{\rho_0}\ \tanh\ y\\ &= \sqrt{\frac{\lambda}{2}}\rho_0\ \frac{1}{\cosh^2 y}\\ \Rightarrow \frac{1}{2}\phi'^2 &= \frac{\lambda}{4}\rho_0^2\frac{1}{\cosh^4 y},\end{aligned} \tag{13.20}$$

and further,

$$\begin{aligned}\phi^2-\rho_0 &= \rho_0(\tanh^2 y - 1)\\ \Rightarrow\ \frac{1}{4}\lambda(\phi^2-\rho_0)^2 &= \frac{1}{4}\rho_0^2(\tanh^2 y - 1)^2\\ &= \frac{1}{4}\rho_0^2\frac{1}{\cosh^4 y}.\end{aligned} \tag{13.21}$$

This gives the result

$$\begin{aligned} T^{00} &= \frac{1}{4}\lambda\rho_0^2\left(\frac{1}{\cosh^4 y} + \frac{1}{\cosh^2 y}\right) \\ &= \frac{\lambda\rho_0^2}{2\cosh^4 y}, \end{aligned} \tag{13.22}$$

with $y = \sqrt{\frac{\lambda\rho_0}{2}}(x-a)$.

(f) The total energy of the "kink" solution is

$$\begin{aligned} E_{\text{kink}} &= \int_{-\infty}^{\infty} dx T^{00} = \int_{-\infty}^{\infty} dy \frac{dx}{dy} T^{00} \\ &= \int_{-\infty}^{\infty} \sqrt{\frac{2}{\lambda\rho_0}} \frac{\lambda\rho_0^2}{2} \frac{1}{\cosh^4 y} dy \\ &= \sqrt{\frac{\lambda\rho_0^3}{2}} \int_{-\infty}^{\infty} \frac{1}{\cosh^2 y} dy \\ &= \frac{2\sqrt{2}}{3}\sqrt{\lambda\rho_0^3}. \end{aligned} \tag{13.23}$$

(g) Solution when the kink moves with velocity $v$ is

$$\phi(z) = \phi(x - vt). \tag{13.24}$$

We introduce the derivatives of $\phi$:

$$\begin{aligned} \phi' &= \frac{\partial z}{\partial x}\frac{\partial \phi}{\partial z} = \frac{\partial \phi}{\partial z}, \\ \phi'' &= \frac{\partial}{\partial x}\frac{\partial \phi}{\partial z} = \frac{\partial z}{\partial x}\frac{\partial^2 \phi}{\partial z^2} = \frac{\partial^2 \phi}{\partial z^2}, \\ \dot{\phi} &= \frac{\partial z}{\partial t}\frac{\partial \phi}{\partial z} = -v^2\frac{\partial \phi}{\partial z}, \\ \ddot{\phi} &= -v\frac{\partial z}{\partial t}\frac{\partial^2 \phi}{\partial z^2} = v^2\frac{\partial^2 \phi}{\partial z^2}. \end{aligned} \tag{13.25}$$

This gives

$$\begin{aligned} &\ddot{\phi} - \phi'' + \lambda\,\phi\,(\phi^2 - \rho_0) = 0 \\ \Rightarrow\ &(v^2 - 1)\frac{\partial^2 \phi}{\partial z^2} + \lambda\phi(\phi^2 - \rho_0) = 0 \\ \Rightarrow\ &\frac{\partial^2 \phi}{\partial z^2} - \frac{\lambda}{1 - v^2}\phi(\phi^2 - \rho_0) = 0, \end{aligned} \tag{13.26}$$

and the result for the energy function is

$$E(v) = \frac{2\sqrt{2}}{3}\sqrt{\frac{\lambda\rho_0^3}{1-v^2}}. \tag{13.27}$$

Compared to $v = 0$, there is an additional factor:

$$\frac{1}{\sqrt{1-v^2}}, \tag{13.28}$$

which is the standard relativistic factor, where the energy of a moving object increases with the velocity $v$ (or more correctly $v/c$).

### *Problem 13.3*

(a) For small variations, the Lagrangian density has the form

$$\mathcal{L}(\dot{y}, \dot{z}, y', z') = \frac{1}{2}(\mu\dot{y}^2 - \sigma y'^2 + \mu\dot{z}^2 - \sigma z'^2). \tag{13.29}$$

This means that the vibrations take place in both the $y$- and $z$-directions. This is different from the waves on the string in the book, where only one direction, $y$, is included. The Lagrangian is the integral over $x$ of the Lagrangian density,

$$L = \int_0^a dx \left[\frac{1}{2}\mu\left\{\left(\frac{\partial y}{\partial t}\right)^2 + \left(\frac{\partial z}{\partial t}\right)^2\right\} - \frac{1}{2}\sigma\left\{\left(\frac{\partial y}{\partial x}\right)^2 + \left(\frac{\partial z}{\partial x}\right)^2\right\}\right]. \tag{13.30}$$

(b) To determine the Euler–Lagrange equations, and solve them, we follow the procedure of the corresponding example in the book. The variation of the the string is as follows:

$$\begin{aligned}\delta S = \int dx \int dt \Bigg[&\left(\frac{\partial}{\partial x}\left(\frac{\partial \mathcal{L}}{\partial y'}\right) + \frac{\partial}{\partial x}\left(\frac{\partial \mathcal{L}}{\partial \dot{y}}\right)\delta y\right. \\ &+ \left(\frac{\partial}{\partial x}\left(\frac{\partial \mathcal{L}}{\partial z'}\right) + \frac{\partial}{\partial x}\left(\frac{\partial \mathcal{L}}{\partial \dot{z}}\right)\delta z\right],\end{aligned} \tag{13.31}$$

where the boundary contribution to $\delta S$ has been set to zero. This is correct for any choice of boundary in the $(x,t)$-plane. The Euler–Lagrange equations are then

$$\frac{\partial}{\partial x}\left(\frac{\partial \mathcal{L}}{\partial y'}\right) + \frac{\partial}{\partial t}\left(\frac{\partial \mathcal{L}}{\partial \dot{y}}\right) = 0, \quad \frac{\partial}{\partial x}\left(\frac{\partial \mathcal{L}}{\partial z'}\right) + \frac{\partial}{\partial t}\left(\frac{\partial \mathcal{L}}{\partial \dot{z}}\right) = 0, \tag{13.32}$$

where

$$\frac{\partial}{\partial t}\left(\frac{\partial \mathcal{L}}{\partial \dot{y}}\right) = \mu \ddot{y}, \quad \frac{\partial}{\partial x}\left(\frac{\partial \mathcal{L}}{\partial y'}\right) = -\sigma y'',$$
$$\frac{\partial}{\partial t}\left(\frac{\partial \mathcal{L}}{\partial \dot{z}}\right) = \mu \ddot{z}, \quad \frac{\partial}{\partial x}\left(\frac{\partial \mathcal{L}}{\partial z'}\right) = -\sigma z''. \tag{13.33}$$

This gives the following equations:

$$\mu \ddot{y} - \sigma y'' = 0 \quad \Rightarrow \quad \frac{\partial^2 y}{\partial t^2} - v^2 \frac{\partial^2 y}{\partial x^2} = 0,$$
$$\mu \ddot{z} - \sigma z'' = 0 \quad \Rightarrow \quad \frac{\partial^2 z}{\partial t^2} - v^2 \frac{\partial^2 z}{\partial x^2} = 0, \tag{13.34}$$

with $v = \sqrt{\sigma/\mu}$.

General solutions are

$$y(x,t) = g(vt + x) - g(vt - x),$$
$$z(x,t) = f(vt + x) - f(vt - x). \tag{13.35}$$

(c) The Fourier expansion expressed with $g(\xi)$ is (see the main book)

$$g(\xi) = \sum_{n=1}^{\infty} \left\{ a_n \cos\left(\frac{n\pi\xi}{a}\right) + b_n \sin\left(\frac{n\pi\xi}{a}\right) \right\}, \tag{13.36}$$

where $x \epsilon [0, a]$. This implies that

$$\begin{aligned} y(x,t) &= g(vt + x) - g(vt - x) \\ &= \sum_n \left[ a_n \left\{ \cos\left(\frac{n\pi}{a}(vt + x)\right) - \cos\left(\frac{n\pi}{a}(vt - x)\right) \right\} \right. \\ &\quad \left. + b_n \left\{ \sin\left(\frac{n\pi}{a}(vt + x)\right) - \sin\left(\frac{n\pi}{a}(vt - x)\right) \right\} \right] \\ &= 2 \sum_n \left[ a_n \left\{ -\sin\left(\frac{n\pi}{a}vt\right) \sin\left(\frac{n\pi}{a}x\right) \right\} \right. \\ &\quad \left. + b_n \left\{ \cos\left(\frac{n\pi}{a}vt\right) \sin\left(\frac{n\pi}{a}x\right) \right\} \right]. \end{aligned} \tag{13.37}$$

We continue with derivatives

$$\begin{aligned} \frac{\partial y}{\partial t} = \sum_n \frac{2n\pi}{a} v &\left( a_n \left\{ -\cos\left(\frac{n\pi}{a}vt\right) \sin\left(\frac{n\pi}{a}x\right) \right\} \right. \\ &\left. - b_n \left\{ \sin\left(\frac{n\pi}{a}vt\right) \sin\left(\frac{n\pi}{a}x\right) \right\} \right), \end{aligned} \tag{13.38}$$

and further,

$$\int_0^a \left(\frac{\partial y}{\partial t}\right)^2 dx = \sum_1^\infty \frac{2n^2\pi^2}{a} v^2 \left(a_n^2 \cos^2\left(\frac{n\pi}{a}vt\right) - b_n^2 \sin^2\left(\frac{n\pi}{a}vt\right) + 2a_n b_n \cos\left(\frac{n\pi}{a}vt\right) \sin\left(\frac{n\pi}{a}vt\right)\right). \quad (13.39)$$

With $\frac{\partial y}{\partial t}$ changed to $\frac{\partial y}{\partial x}$, we get

$$\int_0^a \left(\frac{\partial y}{\partial x}\right)^2 dx = \sum_1^\infty \frac{2n^2\pi^2}{a} v^2 \left(a_n^2 \sin^2\left(\frac{n\pi}{a}vt\right) + b_n^2 \cos^2\left(\frac{n\pi}{a}vt\right) - 2a_n b_n \sin\left(\frac{n\pi}{a}vt\right) \cos\left(\frac{n\pi}{a}vt\right)\right), \quad (13.40)$$

and with the two combined,

$$\int_0^a \frac{1}{2}\mu \left\{\left(\frac{\partial y}{\partial t}\right)^2 - v^2 \left(\frac{\partial y}{\partial x}\right)^2\right\} dx = \sum_1^\infty \frac{n^2\pi^2}{a} \sigma \left(a_n^2 + b_n^2 + 2a_n b_n \sin\left(\frac{n\pi}{a}vt\right)\right). \quad (13.41)$$

A similar result, when $y$ is changed to $z$, is

$$\int_0^a \frac{1}{2}\mu \left\{\left(\frac{\partial z}{\partial t}\right)^2 - v^2 \left(\frac{\partial z}{\partial x}\right)^2\right\} dx = \sum_1^\infty \frac{n^2\pi^2}{a} \sigma \left(c_n^2 + d_n^2 + 2c_n d_n \sin\left(\frac{n\pi}{a}vt\right)\right). \quad (13.42)$$

This gives a form of the Lagrangian density in terms of a Fourier expansion,

$$L = \sum_1^\infty \frac{n^2\pi^2}{a} \sigma \left\{A + 2B \sin\left(\frac{2n\pi}{a}vt\right)\right\}, \quad (13.43)$$

with

$$A^2 = a_n^2 + b_n^2 + c_n^2 + d_n^2, \quad B^2 = a_n b_n + c_n d_n. \quad (13.44)$$

(d) The next points are the Euler–Lagrange equations expressed in terms of the Fourier transform. We start with (13.37):

$$y(x,t) = 2\sum_n \left[a_n \left\{-\sin\left(\frac{n\pi}{a}vt\right) \sin\left(\frac{n\pi}{a}x\right)\right\} + b_n \left\{\cos\left(\frac{n\pi}{a}vt\right) \sin\left(\frac{n\pi}{a}x\right)\right\}\right], \quad (13.45)$$

and we find the derivatives,

$$\frac{\partial^2 y}{\partial t^2} = \sum_n \frac{2n^2\pi^2}{a^2} v^2 \left( a_n \sin\left(\frac{n\pi}{a} vt\right) - b_n \cos \frac{n\pi}{a} vt \right)\right) \sin\left(\frac{n\pi}{a} x\right),$$

$$\frac{\partial^2 y}{\partial x^2} = \sum_n \frac{2n^2\pi^2}{a^2} \left( a_n \sin\left(\frac{n\pi}{a} vt\right) - b_n \cos \frac{n\pi}{a} vt \right)\right) \sin\left(\frac{n\pi}{a} x\right). \quad (13.46)$$

The result is

$$\frac{\partial^2 y}{\partial t^2} - v^2 \frac{\partial^2 y}{\partial x^2} = 0, \quad (13.47)$$

which confirms that the correct Euler–Lagrange equation can also be found with the Fourier transform.

# Chapter 14

# Symmetry transformations

***Problem 14.1***

The Lagranian density,

$$\mathcal{L} = \frac{1}{2}\mu\left\{\left(\frac{\partial y}{\partial t}\right)^2 + \left(\frac{\partial z}{\partial t}\right)^2\right\} - \frac{1}{2}\sigma\left\{\left(\frac{\partial y}{\partial x}\right)^2 + \left(\frac{\partial z}{\partial x}\right)^2\right\}, \quad (14.1)$$

corresponds to the (integrated) Lagrangian in Problem 1.3. The Euler–Lagrange equations here are therefore the same as in that case:

$$\frac{\partial}{\partial x}\left(\frac{\partial \mathcal{L}}{\partial y'}\right) + \frac{\partial}{\partial t}\left(\frac{\partial \mathcal{L}}{\partial \dot{y}}\right) = 0, \quad \frac{\partial}{\partial x}\left(\frac{\partial \mathcal{L}}{\partial z'}\right) + \frac{\partial}{\partial t}\left(\frac{\partial \mathcal{L}}{\partial \dot{z}}\right) = 0. \quad (14.2)$$

As shown there, the equations can be expressed as:

$$\frac{\partial^2 y}{\partial t^2} - v^2\frac{\partial^2 y}{\partial x^2} = 0, \quad \frac{\partial^2 z}{\partial t^2} - v^2\frac{\partial^2 z}{\partial x^2} = 0, \quad (14.3)$$

where $v = \sqrt{\sigma/\mu}$.

(a) The energy–momentum tensor has the general form

$$\begin{aligned} T^{00} &= -\mathcal{L} + \frac{\partial \mathcal{L}}{\partial \dot{y}}y' - \frac{\partial \mathcal{L}}{\partial \dot{z}}z' \\ &= \frac{1}{2}\mu\left\{\left(\frac{\partial y}{\partial t}\right)^2 + \left(\frac{\partial z}{\partial t}\right)^2\right\} + \frac{1}{2}\sigma\left\{\left(\frac{\partial y}{\partial x}\right)^2 + \left(\frac{\partial z}{\partial x}\right)^2\right\} \\ &= T^{11} \end{aligned} \quad (14.4)$$

and

$$\begin{aligned} T^{01} &= -\frac{\partial \mathcal{L}}{\partial \dot{y}}y' - \frac{\partial \mathcal{L}}{\partial \dot{z}}z' \\ &= -\mu\left(\frac{\partial y}{\partial t}\frac{\partial y}{\partial x} + \frac{\partial z}{\partial t}\frac{\partial z}{\partial x}\right) \\ &= T^{10}. \end{aligned} \quad (14.5)$$

The question now is: Are the energy and the momentum conserved?

We begin with the energy density, $\mathcal{E} = T^{00}$. It can be split in two: $\mathcal{E} = \mathcal{E}_y + \mathcal{E}_z$,

$$\mathcal{E}_y = \frac{1}{2}\left\{\mu\left(\frac{\partial y}{\partial t}\right)^2 + \sigma\left(\frac{\partial y}{\partial x}\right)^2\right\}, \quad \mathcal{E}_z = \frac{1}{2}\left\{\mu\left(\frac{\partial z}{\partial t}\right)^2 + \sigma\left(\frac{\partial z}{\partial x}\right)^2\right\}. \tag{14.6}$$

Since these two are equal and independent, it is sufficient to examine one of them. We take $\mathcal{E}_y$, and change the form of this density by use of the corresponding part of the Euler–Lagrange equation,

$$\frac{\partial^2 y}{\partial t^2} - v^2\frac{\partial^2 y}{\partial x^2} = 0, \quad v^2 = \frac{\sigma}{\mu}. \tag{14.7}$$

This implies that for the time derivative of $\mathcal{E}_y$,

$$\begin{aligned}
\frac{\partial \mathcal{E}_y}{\partial t} &= \mu\frac{\partial y}{\partial t}\frac{\partial^2 y}{\partial t^2} + \sigma\frac{\partial y}{\partial x}\frac{\partial^2 y}{\partial t \partial x} \\
&= \sigma\left\{\frac{\partial y}{\partial t}\frac{\partial^2 y}{\partial x^2} + \frac{\partial y}{\partial x}\frac{\partial^2 y}{\partial t \partial x}\right\} \\
&= \sigma\frac{\partial}{\partial x}\left(\frac{\partial y}{\partial t}\frac{\partial y}{\partial x}\right) \\
&= -\frac{\partial j_y}{\partial x}.
\end{aligned} \tag{14.8}$$

The derivative of $\mathcal{E}_z$ gives the same result, with y replaced by z, and the sum of $\mathcal{E}_y$ and $\mathcal{E}_x$, gives the same result:

$$\frac{\partial \mathcal{E}}{\partial t} + \frac{\partial j}{\partial x} = 0, \quad j = j_t + j_x = -\sigma\frac{\partial^2 y}{\partial t \partial x}. \tag{14.9}$$

This result is an interpretation of the conservation of energy, where the change of local energy is compensated by the flow of energy to and from this locality.

$T^{11} = T^{00}$ shows that the flux density appears in the same way as the energy density.

(b) We show that the following transformation is a symmetry transformation:

$$\begin{pmatrix} y \\ z \end{pmatrix} \to \begin{pmatrix} \bar{y} \\ \bar{z} \end{pmatrix} = \begin{pmatrix} \cos\alpha & -\sin\alpha \\ \sin\alpha & \cos\alpha \end{pmatrix}\begin{pmatrix} y \\ z \end{pmatrix}, \tag{14.10}$$

with $\alpha$ as an arbitrary constant.

We note first that

$$\bar{y} = \cos\alpha y - \sin\alpha z,$$
$$\bar{z} = \sin\alpha y + \cos\alpha z, \tag{14.11}$$

and the result under differential of time $t$,

$$\frac{\partial \bar{y}}{\partial t} = \cos\alpha\, \frac{\partial y}{\partial t} - \sin\alpha\, \frac{\partial z}{\partial t},$$
$$\frac{\partial \bar{z}}{\partial t} = \sin\alpha\, \frac{\partial y}{\partial t} + \cos\alpha\, \frac{\partial z}{\partial t}. \tag{14.12}$$

A step further gives

$$\left(\frac{\partial \bar{y}}{\partial t}\right)^2 = \cos^2\alpha \left(\frac{\partial y}{\partial t}\right)^2 - 2\cos\alpha\sin\alpha \frac{\partial y}{\partial t}\frac{\partial z}{\partial t} + \sin^2\alpha \left(\frac{\partial z}{\partial t}\right)^2,$$
$$\left(\frac{\partial \bar{z}}{\partial t}\right)^2 = \sin^2\alpha \left(\frac{\partial y}{\partial t}\right)^2 + 2\cos\alpha\sin\alpha \frac{\partial y}{\partial t}\frac{\partial z}{\partial t} + \cos^2\alpha \left(\frac{\partial z}{\partial t}\right)^2, \tag{14.13}$$

with the result,

$$\left(\frac{\partial \bar{y}}{\partial t}\right)^2 + \left(\frac{\partial \bar{z}}{\partial t}\right)^2 = \left(\frac{\partial y}{\partial t}\right)^2 + \left(\frac{\partial z}{\partial t}\right)^2. \tag{14.14}$$

The same follows if instead of differentiation with respect to $t$, one does that with $x$:

$$\left(\frac{\partial \bar{y}}{\partial x}\right)^2 + \left(\frac{\partial \bar{z}}{\partial x}\right)^2 = \left(\frac{\partial y}{\partial x}\right)^2 + \left(\frac{\partial z}{\partial x}\right)^2. \tag{14.15}$$

This shows that there is a symmetry transformation from $(y, z)$ to $(\bar{y}, \bar{z})$.

The physical meaning is that the Lagrangian density of the string is conserved under rotations of the coordinates $y$ and $z$.

### *Problem 14.2*

A Lagrangian density,

$$\mathcal{L} = -(\phi^{*,\nu}\phi_{,\nu} + \mu^2\phi^*\phi), \tag{14.16}$$

is invariant under the transformation

$$\phi \to \psi = \phi\, e^{i\lambda}. \tag{14.17}$$

(a) We consider the situation where $\lambda$ is infinitesimal:

$$\delta\phi = i\lambda\phi, \quad \delta\phi^* = i\lambda\phi^*. \tag{14.18}$$

The function $N^\mu$, which defines the Noether current, in the present case it takes the form,

$$\begin{aligned} N^\mu &= \frac{\partial \mathcal{L}}{\partial \phi_{,\mu}}\delta\phi + \frac{\partial \mathcal{L}}{\partial \phi^*_{,\mu}}\delta\phi^* \\ &= -i\lambda(\phi^{*,\mu}\phi - \phi^{,\mu}\phi^*). \end{aligned} \tag{14.19}$$

(b) The above expression for $N^\mu$ gives the following derivative:

$$\frac{dN^\mu}{dx^\mu} = -i\lambda\left\{(\partial_\mu\partial^\mu\phi^*)\phi - (\partial_\mu\partial^\mu\phi)\phi^*\right\}, \tag{14.20}$$

and by use of the Euler–Lagrange equation,

$$\begin{aligned} &\frac{\partial \mathcal{L}}{\partial \phi} - \frac{d}{dx^\mu}\left(\frac{\partial \mathcal{L}}{\partial \phi_{,\mu}}\right) = 0 \\ \Rightarrow \quad &(\Box - \mu^2)\phi^* = 0, \quad (\Box - \mu^2)\phi = 0, \end{aligned} \tag{14.21}$$

we find

$$\frac{dN^\mu}{dx^\mu} = -i\lambda\mu^2\left\{\phi^*\phi - \phi\phi^*\right\} = 0, \tag{14.22}$$

which means that the conservation law is satisfied:

$$\frac{dN^\mu}{dx^\mu} = 0. \tag{14.23}$$

(c) To find the energy–momentum density tensor, $T^{\mu\nu}$, in this case, is as follows:

$$\begin{aligned} T^{\mu\nu} &= g^{\mu\nu}\mathcal{L} - \frac{\partial \mathcal{L}}{\partial \phi_{,\mu}}\phi^{,\nu} - \frac{\partial \mathcal{L}}{\partial \phi^*_{,\mu}}\phi^{*,\nu} \\ &= -g^{\mu\nu}(\phi^{*,\lambda}\phi_{,\lambda} + \mu^2\phi^*\phi) + \phi^{*,\mu}\phi^{,\nu} + \phi^{*,\nu}\phi^{,\mu}. \end{aligned} \tag{14.24}$$

We will look at the components, first the energy density,

$$\begin{aligned} T^{00} &= \phi^{*,0}\phi^{,0} + \phi^{*,k}\phi^{,k} + \mu^2\phi^*\phi \\ &= |\dot\phi|^2 + |\phi^{,k}|^2 + \mu^2|\phi|^2, \end{aligned} \tag{14.25}$$

and next the flux density in the direction of $x^j$,

$$\begin{aligned} T^{j0} &= -\frac{\partial \mathcal{L}}{\partial \phi_{,j}}\phi^{,0} - \frac{\partial \mathcal{L}}{\partial \phi^*_{,j}}\phi^{*,0} \\ &= \phi^{*,j}\phi^{,0} + \phi^{*,0}\phi^{,j}. \end{aligned} \tag{14.26}$$

The $k$-momentum density is

$$T^{0k} = -\frac{\partial \mathcal{L}}{\partial \phi_{,0}}\phi^{,k} - \frac{\partial \mathcal{L}}{\partial \phi^*_{,0}}\phi^{*,k}$$
$$= \phi^{*,k}\phi^{,0} + \phi^{*,0}\phi^{,k}, \tag{14.27}$$

and finally, the flux density of $k$-momentum in the direction $x^j$,

$$T^{jk} = -\delta_{jk}(\phi^{*,\lambda}\phi_{,\lambda} + \mu^2\phi^*\phi) + \phi^{*,j}\phi^{,k} + \phi^{*,k}\phi^{,j}. \tag{14.28}$$

### *Problem 14.3*

(a) We study the the following Lagrangian density:

$$\mathcal{L} = \frac{1}{2}i\hbar(\phi^*\dot{\phi} - \dot{\phi}^*\phi) - \frac{\hbar^2}{2m}\nabla\phi^* \cdot \nabla\phi - V\phi^*\phi, \tag{14.29}$$

where $\phi = \phi(\mathbf{r},t)$ and $V = V(\mathbf{r},t)$. The problem is to show that Euler–Lagrange equations will have the form of a Schrödinger equation and its complex conjugate.

The Euler–Lagrange equations are

$$\frac{\partial \mathcal{L}}{\partial \phi^*} - \frac{d}{dx^\mu}\left(\frac{\partial \mathcal{L}}{\partial \phi^*_{,\mu}}\right) = 0 \tag{14.30}$$

and the complex conjugate. We look at the components of the equation:

$$\begin{aligned}
\frac{\partial \mathcal{L}}{\partial \phi^*} &= \frac{1}{2}i\hbar\dot{\phi} + V\phi, \\
\frac{\partial \mathcal{L}}{\partial \phi^*_{,t}} &= -i\hbar\phi \quad \Rightarrow \quad \frac{d}{dt}\frac{\partial \mathcal{L}}{\partial \phi} = -\frac{1}{2}i\hbar\dot{\phi}, \\
\frac{\partial \mathcal{L}}{\partial \phi^*_{,k}} &= -\frac{\hbar^2}{2m}\frac{\partial \phi}{\partial x^{,k}} \quad \Rightarrow \quad \frac{d}{dx^k}\frac{\partial \mathcal{L}}{\partial \phi^*_{,k}} = -\frac{\hbar^2}{2m}\frac{\partial^2 \phi}{\partial x^k dx_k}.
\end{aligned} \tag{14.31}$$

This gives for the Euler–Lagrange equation,

$$\frac{1}{2}i\hbar\dot{\phi} + \frac{1}{2}i\hbar\dot{\phi} + \frac{\hbar^2}{2m}\frac{\partial^2\phi}{\partial x^k dx_k} = 0$$
$$\Rightarrow \quad i\hbar\frac{\partial\phi}{\partial t} = -\frac{\partial^2\phi}{\partial x^k dx_k} + V(\mathbf{r},t). \tag{14.32}$$

This has the same form as the Schrödinger equation. There is also the complex conjugate of the equation.

b) The condition for symmetry under time translation is that the transformation has the form,

$$\psi(y) = \phi(x), \quad with \quad y^\mu = x^\mu + b\,\delta_0^\mu. \tag{14.33}$$

The corresponding conserved quantity can be expressed as

$$\frac{dT^{\mu 0}}{dx^\mu} = 0. \tag{14.34}$$

This means that the energy–momentum density is conserved in the time direction.

The condition for symmetry under a space translation (in the $k$-direction) means that the transformation can now be written as

$$\frac{dT^{\mu k}}{dx^\mu} = 0. \tag{14.35}$$

This means that the energy–momentum density is conserved in the $k$-direction.

# Chapter 15

# Relativistic fields

*Problem 15.1*

We have in this problem the follow Lagrangian density:

$$\mathcal{L} = -\frac{1}{2}\partial_\mu \phi \partial^\mu \phi + \alpha(\cos\phi - 1), \tag{15.1}$$

with $\alpha$ as a real constant.

(a) The Euler–Lagrange equation is

$$\frac{\partial \mathcal{L}}{\partial \phi} - \frac{d}{dx^\mu}\left(\frac{\partial \mathcal{L}}{\partial \phi_{,\mu}}\right) = 0 \quad \rightarrow \quad \alpha \sin\phi - \partial_\mu \partial^\mu \phi = 0, \tag{15.2}$$

and the following two-dimensional solution is given as

$$\phi(x,t) = 4 \arctan\left[\frac{\beta \sinh(\sqrt{\alpha}\gamma x)}{\cosh(\sqrt{\alpha}\gamma v t)}\right], \tag{15.3}$$

with $\beta = v/c$ and $\gamma = 1/\sqrt{1-\beta^2}$.

(b) We study the time evolution of a collision process by plotting $\phi$ as a function of $x$ for a series of time $t$. There are two series with different velocities. In one set, (a), the velocity $v$ is slow, $\beta = 0.1$, and in the other set, (b), the velocity $v$ is large, $\beta = 0.9$. The interpretation of the motion of the kinks is that they first move toward the centre $x = 0$, where they stop and then move in the opposite direction. In collision (a), the centre of the kink seems to stop and returns earlier than the kink in collision (b). This is consistent with the situation where the collision (a) moves slowly and collision (b) moves fast. It is also consistent with the solitons being repulsive.

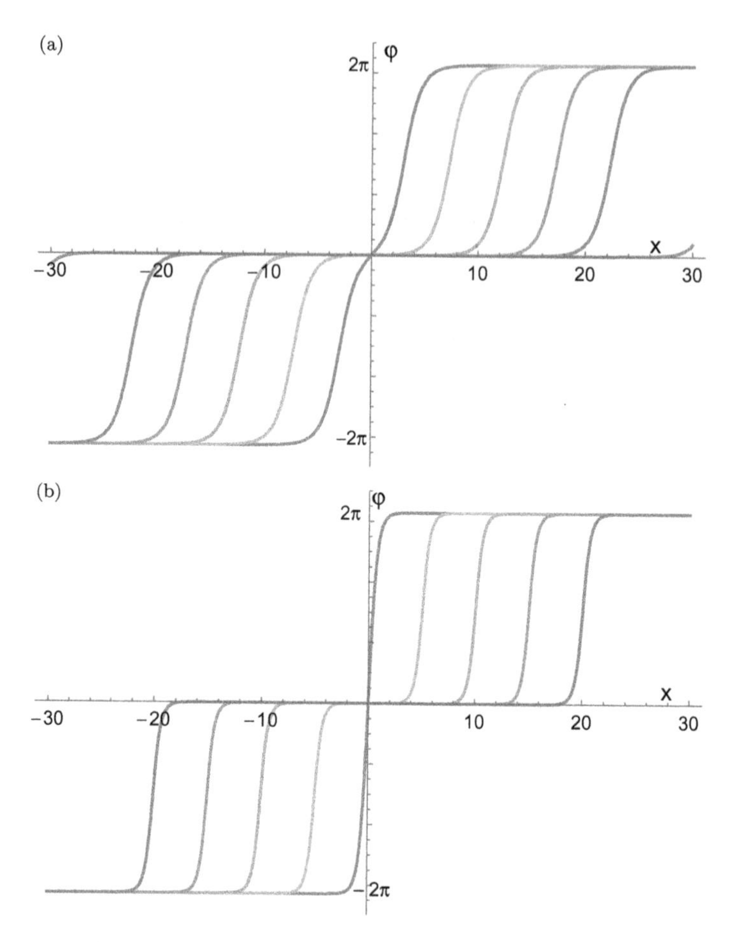


Fig. 15.1 Collisions between two kink solitons. There are two different types of collisions, where (a) is a slow collision and (b) is a rapid collision. The different colored lines show a sequence of positions of the solitons at increasing times. Since the solitons before and after the collision are precisely the same, both the situations before and after the collisions can be seen here.

## *Problem 15.2*

(a) The Lagrangian density for the electromagnetic field with an extra term added is

$$\begin{aligned}\mathcal{L} &= -\frac{1}{4\mu_0}\left(F^{\mu\nu}F_{\mu\nu} - 2\,\partial^\mu A_\mu\,\partial^\nu A_\nu\right) + j_\mu A^\mu \\ &= -\frac{1}{2\mu_0}\left(\partial^\nu A^\mu\,\partial_\nu A_\mu - \partial^\nu A^\mu\,\partial_\nu A_\mu - \partial^\mu A_\mu\,\partial^\nu A_\nu\right) + j_\mu A^\mu. \quad (15.4)\end{aligned}$$

The Euler–Lagrange equation is

$$\frac{\partial \mathcal{L}}{\partial A^\mu} - \frac{d}{dx^\nu}\left(\frac{\partial \mathcal{L}}{\partial A^\mu_{,\,\nu}}\right) = 0,$$

$$\Rightarrow \quad \frac{\partial \mathcal{L}}{\partial A^\mu} = j^\mu,$$

$$\frac{\partial \mathcal{L}}{\partial A^\mu{}_{,\nu}} = -\frac{1}{\mu_0}\left(A_\mu^{;\nu} - A^\nu{}_{,\mu} + \frac{1}{2}\frac{\partial}{\partial A^\mu{}_{,\nu}} A^\alpha{}_{,\alpha} A^\beta{}_{,\beta}\right). \tag{15.5}$$

This gives

$$\begin{aligned}
\partial_\nu\left(\frac{\partial \mathcal{L}}{\partial A^\mu{}_{,\nu}}\right) &= -\frac{1}{\mu_0}\left(\partial_\nu\partial^\nu A_\mu - \partial_\nu\partial_\mu A^\nu + \partial_\nu \frac{\partial A^\alpha{}_{,\alpha}}{\partial A^\mu{}_{,\nu}} A^\beta{}_{,\beta}\right)\\
&= -\frac{1}{\mu_0}\left(\partial_\nu\partial^\nu A_\mu - \partial_\nu\partial_\mu A^\nu + \partial_\mu\partial_\nu A^\nu\right)\\
&= -\frac{1}{\mu_0}\partial_\nu\partial^\nu A_\mu.
\end{aligned} \tag{15.6}$$

The full result for the Euler–Lagrange equation is then

$$\Box A^\mu = -\mu_0 j^\mu. \tag{15.7}$$

(b) This form of the equation is different from which we have met before:

$$\Box A^\mu - \frac{\partial}{\partial x^\mu}\left(\frac{\partial A^\nu}{\partial x^\nu}\right) = -\mu_0 j^\mu. \tag{15.8}$$

The difference between the two descriptions of the Euler–Lagrange equations is a gauge transformation, where the difference has the form,

$$\partial^\mu \chi(x) = \partial^\mu \partial_\nu A^\nu(x). \tag{15.9}$$

### *Problem 15.3*

The Lagrange density of the complex Klein–Gordon field which interacts with a Maxwell field are

$$\begin{aligned}
&\mathcal{L}(A_\mu, A_{\mu,\nu}; \phi, \phi_{,\mu}, \phi^*_{,}\phi^*_{,\mu})\\
&= -\frac{1}{4\mu_0}F_{\mu\nu}F^{\mu\nu} - \left(\left[(\partial_\nu - i\frac{q}{\hbar c}A_\nu)\phi\right]^* \left[(\partial_\nu - i\frac{q}{\hbar c}A^\nu)\phi\right] + \mu^2\phi^*\phi\right)
\end{aligned}$$

$$= -\frac{1}{2\mu_0}(A_{\nu,\mu}\, A^{\nu,\mu} - A_{\nu,\mu}A^{\mu,\nu}) - \frac{q^2}{\hbar^2 c^2}A_\nu A^\nu \phi^* \phi$$

$$+ i\frac{q}{\hbar c}A^\nu(\phi^*_{,\nu}\phi - \phi_{,\nu}\phi^*) - (\phi^*_{,\nu}\phi^{,\nu} + \mu^2\phi^*\phi). \tag{15.10}$$

(a) The Euler–Lagrange equations are three equations, with the first one as

$$\frac{\partial \mathcal{L}}{\partial A^\nu} - \frac{d}{dx^\mu}\left(\frac{\partial \mathcal{L}}{\partial A^\nu{}_{,\,\mu}}\right) = 0,$$

$$\Rightarrow \quad \frac{\partial \mathcal{L}}{\partial A^\nu} = -\frac{q^2}{\hbar c^2}A_\nu\ \phi^*\phi + i\frac{q}{\hbar c}(\phi^*_{,\nu}\phi - \phi_{,\nu}\phi^*),$$

$$\frac{d}{dx^\mu}\frac{\partial \mathcal{L}}{\partial A^\mu{}_{,\nu}} = \frac{1}{\mu_0}(\partial_\mu\partial_\nu A^\mu - \partial_\mu\partial^\mu A_\nu). \tag{15.11}$$

The result is

$$\frac{1}{\mu_0}(\partial_\mu\partial_\nu A^\mu - \partial_\mu\partial^\mu A_\nu) - \frac{q^2}{\hbar^2 c^2}A_\nu\phi^*\phi + \frac{q}{\hbar c}(\phi^*_{,\nu}\phi - \phi_{,\nu}\phi^*) = 0. \tag{15.12}$$

The next one is

$$\frac{\partial \mathcal{L}}{\partial \phi^*} - \frac{d}{dx^\mu}\left(\frac{\partial \mathcal{L}}{\partial \phi^*_{,\mu}}\right) = 0,$$

$$\Rightarrow \quad \frac{\partial \mathcal{L}}{\partial \phi^*} = -\left(\frac{q^2}{\hbar^2 c^2}A_\nu A^\nu + \mu^2\right)\phi - i\frac{q}{\hbar c}A^\nu\phi_{,\nu},$$

$$\frac{d}{dx^\nu}\frac{\partial \mathcal{L}}{\partial \phi^*_{,\nu}} = i\frac{q}{\hbar c}(\partial_\nu A^\nu\phi + A^\nu\phi_{,\nu}), \tag{15.13}$$

with the result

$$\left(\frac{q^2}{\hbar^2 c^2}(A_\nu A^\nu + \partial_\nu A^\nu) + \mu^2\right)\phi - i\frac{q}{\hbar c}(\partial_\nu A^\nu\phi + 2A^\nu\phi_{,\nu}) = 0. \tag{15.14}$$

The last one is the same, up to the difference $\phi^* \to \phi$, as in the previous case:

$$\frac{\partial \mathcal{L}}{\partial \phi} - \frac{d}{dx^\mu}\left(\frac{\partial \mathcal{L}}{\partial \phi_{,\mu}}\right) = 0, \tag{15.15}$$

which has the result,

$$\left(\frac{q^2}{\hbar^2 c^2}(A_\nu A^\nu + \partial_\nu A^\nu) + \mu^2\right)\phi^* + i\frac{q}{\hbar c}(\partial_\nu A^\nu\phi + 2A^\nu\phi_\nu) = 0. \tag{15.16}$$

(b) We will now find the Noether current by use of phase transformations og $\phi$ and $\phi^*$. The transformation has the form,

$$\phi \to \phi' = e^{i\chi}\phi, \quad \phi^* \to \phi'^* = e^{-i\chi}\phi^*, \tag{15.17}$$

where $\chi$ is a real number. When $\chi$ is infinitesimal we write it as

$$\delta\phi = i\chi\phi, \quad \delta\phi^* = -i\chi\phi^*. \tag{15.18}$$

The Noether function $N^\mu$ is defined as

$$\begin{aligned} N^\mu &= \frac{\partial \mathcal{L}}{\partial \phi_{,\mu}}\delta\phi + \frac{\partial \mathcal{L}}{\partial \phi^*_{,\mu}}\delta\phi^* \\ &= 2\frac{q}{\hbar c}\chi A^\mu \phi\phi^* - i\chi(\phi^{*,\mu}\phi - \phi^{,\mu}\phi^*). \end{aligned} \tag{15.19}$$

This implies that

$$\partial_\mu N^\mu = 2\frac{q}{\hbar c}\chi\ \partial_\mu A^\mu\ \phi\phi^*, \tag{15.20}$$

and we note that $\partial_\mu A^\mu$ is included. But this is a gauge-dependent quantity, which can be changed to zero. This dependence of $A^\mu$ in $N^\mu$ can thus be changed so that the derivative is zero:

$$\frac{dN^\mu}{dx^\mu} = 0. \tag{15.21}$$

(c) We will show that the following changes are invariant under gauge transformations:

$$\begin{aligned} A_\mu(x) &\to A_\mu + \partial_\mu\theta(x), \\ \phi(x) &\to \exp\left[i\frac{q}{\hbar c}\theta(x)\right]\phi(x), \\ \phi^*(x) &\to \exp\left[-i\frac{q}{\hbar c}\theta(x)\right]\phi^*(x). \end{aligned} \tag{15.22}$$

To show this, we study how the Lagrangian density is affected by the changes.

We first look at how the following part of the Lagrangian density is changed:

$$\begin{aligned} \left(\partial^\nu - i\frac{g}{\hbar c}A^\nu\right)\phi &\to \left(\partial^\nu - i\frac{g}{\hbar c}(A^\nu + \partial_\mu\theta(x)\right)\exp\left[i\frac{q}{\hbar c}\theta(x)\right]\phi(x) \\ &= \exp\left[i\frac{q}{\hbar c}\theta(x)\right]\left(\partial^\nu - i\frac{q}{\hbar c}A^\nu\right)\phi(x). \end{aligned} \tag{15.23}$$

We see the result that the exponential part is moved, and the term $\partial_\mu \theta(x)$ is removed. This effect disappears altogether when the conjugate part is introduced as follows:

$$\left[\left(\partial_\nu - i\frac{q}{\hbar c}A_\nu\right)\phi\right]^* \left[\left(\partial_\nu - i\frac{q}{\hbar c}A^\nu\right),\phi\right]. \tag{15.24}$$

Also, the field equations are invariant under the gauge transformations, since these are extracted from the Lagrangian density.

www.ingramcontent.com/pod-product-compliance
Lightning Source LLC
LaVergne TN
LVHW022022260125
802164LV00001B/13

* 9 7 8 9 8 1 9 8 0 7 5 2 9 *